THE ALL IS AN EGG

A SYNTHETIC THEORY OF THE UNIVERSE, HUMANKIND AND RELIGION

By Urban Vyaas

BOSTOEN
COPELAND & DAY
2020

For more information, address: urban.vyaas@gmail.com.

Publishers: Bostoen, Copeland and Day, Wönnichstrasse 110, 10317 Berlin
First paperback edition
ISBN 9798651952328

Cover design; Metamorphosis of Narcissus by Salvador Dali (1934)

Introduction.

If you´re looking for an easy read, then this book might not be for you. It is also not for the scientists who have forgotten all their high school knowledge that´s not connected with their field of expertise or the artists who believe that they can create something meaningful by unlearning all the knowledge they have accumulated during their formative years.
This book has been created for the life-long learners who crave to "think-the-world-together" again. The required preexisting knowledge rarely surmounts the level of the subjects that are taught at high school.
While it is a generally accepted adage that new paradigms in science, art and religion are often meeting active resistance from the people who adhere to the commonly accepted views and methods, they're nevertheless always new insights and perspectives budding up from the old canon.
The inspiration for this essay occurred to me after a visit of the Dali museum in Figueres at a time when I was contemplating about a new concept that could reunite again the fragmented field of human knowledge and skills. And there I stumbled upon the egg.

Figure 1: Dali Theater-museum in Figueres

The egg is a dominant theme in Dali´s paintings and sculptures and I came spontaneously to wonder; why can the eggs be found everywhere in Dali's works?

One possible explanation is that the egg is intrinsically linked to the birth of a new life.

In a more prosaic way, we can also see in this fascination with the egg, a manifestation of his Catalan spirit. In Barcelona and the surrounding area the egg is eaten everywhere: as a snack in the pub up to become an elaborate dish in the gastronomical restaurants. And then you have its intriguing topological aspects.

The factor that characterizes eggs of all sizes is that all of them are obtained by turning a flat curve around its axis. The complication is caused by the fact that the sections made with planes perpendicular to the axis are all circumferences of different sizes (which tells us that the egg it is a rotation surface), those obtained by cutting with planes containing the vertical axis are neither circumferences nor ellipses, but strange curves in which a half is more paunchy than the other. It´s oval shape has also intrigued scientists as Cassini, Descartes, Kepler, Granville, Hügelschäffer and Moss.

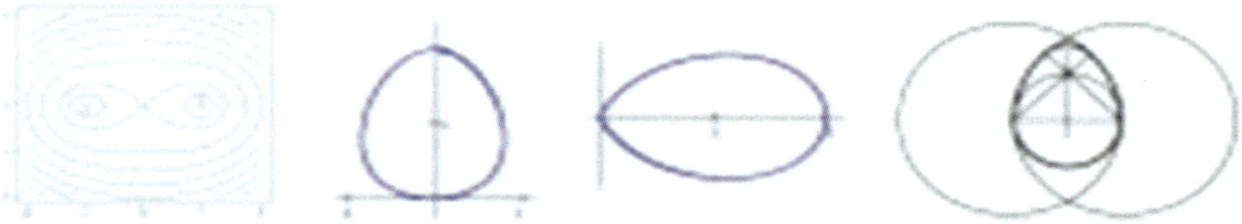

Figure 2 **Mathematical Eggs: Cassini, Granville, Kepler and Moss.**

Mathematics are not only a good technical support for artists, but also an additional creative stimulus. The common opinion tends to be rather hostile to such an assertion. Analyzing Salvador Dali's work, on the contrary, is a great way to support such a daring thesis. His book *50 Secrets of Magic Craftsmanship*, contains the following advice:

"You must, especially as a young man, use geometry as a guide to symmetry in the composition of your works. I know that more or less romantic painters argue that these mathematical scaffolds kill the

artist's inspiration, giving him too much to think and reflect. Do not hesitate for a moment to respond promptly that, on the contrary, it is just not to have to think and reflect on certain things that you use them".

His "recipe" for beauty was to put close geometric constraints at the base of a picture, and then to let his creativity flow, sure that the result will be aesthetically harmonic.

Figure 3 *Atomic Leda* and its preparatory study

Leda is one of the examples of paintings based on the concept of *golden section*, foisted upon the geometry of the pentagon. In a regular pentagon, in fact, the relationship between each diagonal and the side is the same as this number, Φ, known in the centuries as the golden ratio (look here to learn more). All artists have been fascinated by this number, and Dalí is conformist in this respect: golden rectangles and related objects pervade his work even when it does not seem.

The Fibonacci sequence (and with it the Golden Ratio) can be seen throughout our universe. From our strands of DNA to the arms of a spiral galaxy, these numbers seem to pop up everywhere. One place that many people don't consider when thinking of the Fibonacci sequence is literature. However, it can be found in writing throughout many time periods and cultures.

In Sanskrit poetry, there are two types of syllables: light (laghu) and heavy (guru). The laghu syllables take up one beat and guru syllables take up two. These syllables can be arranged in different patterns in a line of poetry. After a while, some of the more philosophical poets began to wonder how many arrangements of laghu and guru syllables they could make if they were given a set number of beats. The answer they came up with might look familiar: if a line has n beats, then the number of 1 beat laghu arrangements it can have is n-1 (since n-1+1=n), and it can have n-2 guru patterns (since n-2+2=n). To put it in a general formula, if Pn is the number of arrangements given n number of beats, then Pn=Pn-1+Pn-2. In other words, these poets had figured out the formula for the Fibonnaci sequence before he was even born!

Shakespeare's most common meter was iambic pentameter, which uses a lot of Fibonacci numbers. This meter is a pattern of one unstressed syllable followed by one stressed syllable. This pattern of two syllables is repeated five times per line.

The most common setup for an essay is a five-paragraph essay. (Already the presence of the Fibonacci sequence is present- 5 is a Fibonacci number.) This five-paragraph essay is broken up into smaller parts: ONE intro paragraph, THREE body paragraphs and ONE conclusion paragraph. This leads to there being TWO paragraphs summarizing the given topic and THREE paragraphs actual exploring it. Additionally, an ideal number of sentences in the middle paragraphs is EIGHT, with THREE being the least amount a paragraph can have (normally the conclusion paragraph).

Big parts of the literary imagination align with the egg’s topology, thus affirming the thesis that words and numbers are twin powers that create value in our world. The domains of mathematics and literature complement each other in a variety of ways, even as they attempt to subvert and invert one another.

By re-conceiving the literary work, not as a static, timeless, and ultimately isolated object, but instead as a socially embedded, circulatory process – as an event that can be mapped – topology can help us rethink the nature of the work and literary work itself, and by extension theories of authorship that underpin such work. The mysterious dimensional aspects of our existence are connected to our inhabiting a world that also inhabits us.

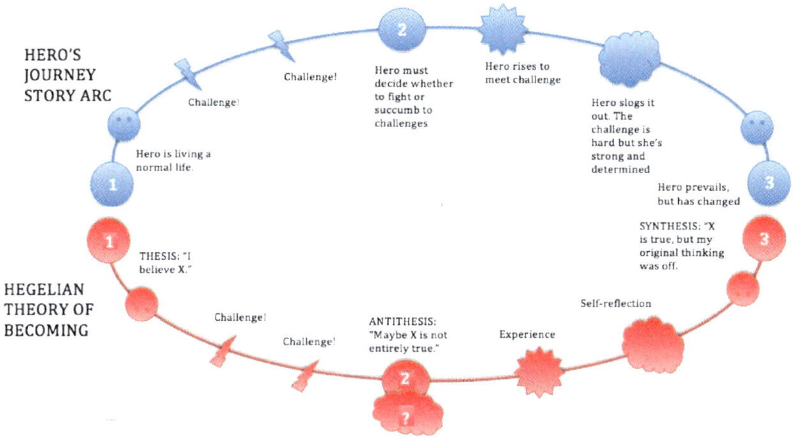
HERO'S JOURNEY STORY ARC
Hero is living a normal life.
Challenge!
Challenge!
Hero must decide whether to fight or succumb to challenges
Hero rises to meet challenge
Hero slogs it out. The challenge is hard but she's strong and determined
Hero prevails, but has changed
1
2
3
HEGELIAN THEORY OF BECOMING
THESIS: "I believe X."
Challenge!
Challenge!
ANTITHESIS: "Maybe X is not entirely true."
Experience
Self-reflection
SYNTHESIS: "X is true, but my original thinking was off.
1
2
?
3

In religion the egg is often used as a symbol for resurrection: a theme that has occupied religious leaders, artists and scientist of all times and continents. In most literary works that use this device, the protagonists are usually not only resurrected, but also given additional powers.

The oldest book that I know of that mentions this literary device is the Mahabharata, where the main protagonists, four of the Pandavas, are resurrected from the dead by Dharma at the end of part three and are given some extra boons that will help them to overcome some future setbacks.

The second book that comes to mind is the world's number one bestseller; the Bible, where not only the main protagonist rises a friend from the dead but also rises from the dead by himself three days after his crucifixion.

The theme is also explored by more contemporary works of literature, where countless heroes have crossed the boundaries of the Earthly realm and come back as a changed person. The only aspect into which the modern variations upon this theme deviate from the traditional tales, is that not all of them come back as improved beings. Some of those resurrections went horribly wrong and the protagonist comes back as a monster or a decomposing corpse.

Examples are Herbert White in The Monkey's Paw, by W.W. Jacobs, Emma Wintertowne in Jonathan Strange & Mr. Norrell, by Susanna Clarke or Ligeia / Rowena in "Ligeia," by Edgar Allan Poe.

Some novels where the resurrection process went well are Gandalf in The Lord of the Rings, by J.R.R. Tolkien, who comes back as the even more powerful Gandalf The White after he died on Zirak Zigil or Aslan in The Lion, the Witch and the Wardrobe, by C.S. Lewis who is by many reviewers seen as an allegory of Christ and who comes back the morning after the White Witch killed him to conquer his enemy.

The return from the grave is also a theme that has been exhaustively explored by many science fiction writers who speculate about new evolutions in biotechnology that would allow cloned people to be resurrected with their memories, intellect and personalities intact. In Dune by Frank Herbert, the Tleilaxu can grow clones from some cells of deceased people, who can recover their genetic memory after a carefully staged moment of extreme stress.

From there on it's a small jump to actual scientific research where engineer and futurist Ian Pearson predicts technology will allow people to upload consciousness to the cloud by 2050. He describes a near future in which we integrate our consciousnesses with technology to the point where our minds use the external space as much as our own brains and bodies. Pearson writes: "One day, your body dies and with it your brain stops, but no big problem, because 99% of your mind is still fine, running happily on information technology, in the cloud."

The grail of eternal life and resurrection isn't anymore the monopoly of theologians, priests and other religious leaders, but has become also a recurring subject of intense research and speculation among artists and scientists.

Big parts of this essay are founded upon the idea how the cyclical nature of a multitude of phenomena mirrors circulatory biological patterns. Connecting literature, philosophy, mathematics, and science, is a transformative work, and a profound invitation to reflect, with the theme of the egg foisted onto it.

Current cosmological models maintain that 13.8 billion years ago, the entire mass of the universe was compressed into a gravitational singularity, the so-called cosmic egg, from which it expanded to its current state.

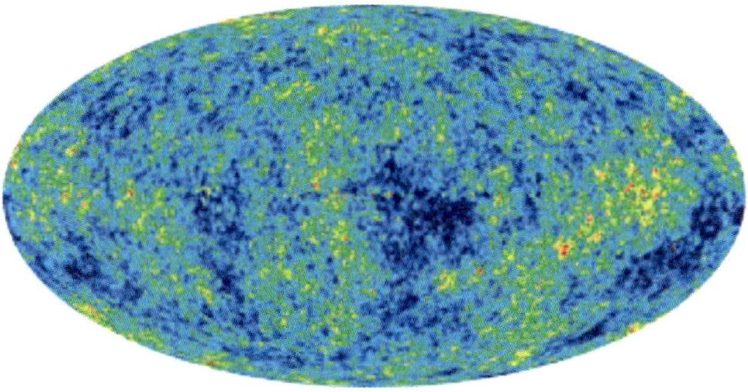

Figure 4: **Shape of the Universe (Lambda CDM model)**

The French paleontologist and philosopher Teilhard de Chardin discerned into his book *The Human Phenomenon*[1] three phases of creation leading to intelligent, self-conscious life on Earth;

- De creation of the Geosphere; water, minerals, air.
- Out of the water came all life; the Biosphere. Noteworthy into this context is the (still ongoing) Miller experiment, started in 1952. It simulated the chemical conditions on the Early Earth and synthesized complex organic compound from their simpler inorganic components by just placing the latter into a huge glass vial filled with water and exposed the content to high electrical currents (simulating weather conditions that occurred at that time).

---

[1] The Human Phenomenon (1999), Brighton: Sussex Academic, 2003: ISBN 1-902210-30-1

- From the Biosphere emerged the Noosphere; intelligence, knowledge and self-consciousness.

However, most critics find the last part of Teilhard's book, where he predicts that the Noosphere would evolve to a maximum level of complexity and consciousness, called the Omega Point, in order to ultimately reunite with Christ, too speculative and probably added in an (vain) effort to gain permission for publication from his ecclesial superiors. Instead, this essay postulates that there exists a future point where religion, art and science might fusion. But before this to happen, there is a need for a coherent frame for all these aspects of human knowledge.
The concept of this essay develops itself along the broad lines Teilhard has designed in an effort to create a unifying frame for the fragmentized particles of the human knowledge.
A first part is dedicated to give the reader an oversight of our actual understanding of the cosmogenesis; the second part presents a mathematical model that unifies our contemporary understanding of the way life came into existence, how it evolves and gave birth to the human brain; the third part offers the reader a blueprint of the way the noosphere might be structured; the fourth part discusses the painful collapse of an old idea of a god to create space for a more refined idea on religion.
This essay is not of an encyclopedic nature, but rather focuses on the mathematically structure of the driving forces behind the reality. If the ideal of the "unity of science", bridging both diverse contents and cultural differences can be achieved at all, this will be done via mathematization.
In the human sciences, religion and art, qualitative parameters have sometimes been used to demonstrate a crypto-mathematically relation with sociologically relevant phenomena, thus allowing their incorporation into the mathematical models that natural sciences have developed. The use of qualitative data to establish synchronic algebraic relations between non-quantitative parameters is a speculative issue since such finalism or teleology is considered to be outside the domain of modern science.
Synchronicity is a widely misunderstood phenomenon and a very often abused term to describe a onetime lucky coincidence or event. Karl Jung defined synchronicity as following; "A meaningful

coincidence of two or more events where something other than the probability of chance is involved. When coincidences pile up in this way, one cannot help being impressed by them – for the greater the number of terms in such a series, or the more unusual its character, the more improbable it becomes." He believed life was not a series of random events but rather an expression of a deeper order, which he and Pauli referred to as Unus mundus. The traditional notions of causality were incapable of explaining some of the more improbable forms of coincidence. Where it is plain, felt Jung that no causal connection can be demonstrated between two events, but where a meaningful relationship nevertheless exists between them, a wholly different type of principle is likely to be operating. Jung called this principle "synchronicity."

Belief and morals always precede logics and are in fact the real foundations of the "Principium Contradictoris"; the foundation upon the whole logical system is based. Something is or isn't; it cannot be both. Gödel [2] proved already that the consistency of any given formal system cannot be proven with its own means. Because we assume that the world is logically consistent, we deduct that contradictions cannot relate to any fact. But paradoxes exist and have always a bigger compelling force than their refutations and are therefore, in a certain sense, more valid.

The modern opinion of logics is that it doesn't relate to the structure of reality but only is valid for the possible languages used to order this reality.

This essay does not try to resolve this paradox, because that's impossible, but wants to revolve with it.

---

[2] Kurt Godel, 1992. On Formally Undecidable Propositions Of Principia Mathematica And Related Systems, tr. B. Meltzer, with a comprehensive introduction by Richard Braithwaite. Dover reprint of the 1962 Basic Books edition.

Table of Content

THE ALL IS AN EGG 1
A SYNTHETIC THEORY OF THE UNIVERSE, HUMANKIND AND RELIGION 1
By Urban Vyaas 1
Introduction. 3
Table of Content 12
Part 1; The GEOSPHERE; What we know and don't know about the Cosmos Genesis. 15
1.1 Einstein's Relativity theory. 15
1.2. The Quantum Mechanics Field Theory 17
1.3. Unresolved Issues in the Grand Unified Theory 22
Part 2; The BIOSPHERE 27
2.1 Definition of the Biosphere 27
2.2. Mathematical Model of the Biosphere; Out of the Water Comes all Life. 28
2.3. The Evolution of Life. 32
2.4. The Human Brain. 35
2.4.1. The Neurophilosophy of Sperry. 35
2.4.2. The Mathematical Neurological Model of Johannes Friedrich and Máté Lengye 38
2.4.3. The Bellman optimization equation in our decision making. 41
4.2.3.1. An illustration of the working of the algorithm 41
4.2.3.2. A formal analysis of the story. 42
4.2.3.3. The Relevance of the Story 43
2.4.3.4. Method 44
2.4.4. Analysis and Discussion. 45
2.4.5. Conclusion 46
Part 3. The Noosphere 48
4.2. The Noosphere and Memetics. 48
3.2. Evolution and Human History. 51
3.2.1. The Mathematical Model of Claudio Maccone. 51
3.2.2. The Inclusion of China into the model. 59
3.3. The Seven-Dimensional Nature of Culture 62
3.4. The Synthetic Approach of Literature. 65
3.4.1. A Multidisciplinary Angle. 65
3.4.2. Literary Criticism and Mathematics 73

3.4.3. A Brief Contextual Synopsis of the US Literary Canon. 76
3.4.4. The Universe and the US Literary System. 87
3.4.5. The US Literary System and the Bible 90
3.4.6. Trend-lines in US literature. 91
3.4.6.1. Methodology 91
3.4.6.2. Trend-lines in US fiction between 1863 -2015......92
3.4.6.3. Protagonist fluctuation in American bestselling Novels between 1863 and 1963. 94
3.4.6.4. Subject periodogram of the American bestselling nonfiction between 1917 -2017. 96
3.4.7. The Multivariate Mathematical Model of North American Literature. 97
3.4.8. Analogies from scientifically Literature. 103
3.4.8.1. The propagation of literary works 103
3.4.8.2. The exploration of literary expansion processes described by differential equations. 103
3.4.8.3. The use of a conical coordinate system. 105
3.4.8.4. Improving the validation of the obtained results. 108
3.4.9. The use of the model to forecast literary trends 110
3.4.10. Indications of unexplored fields in US literary criticism. 111
3.5. What about Awareness? 115
3.5.1. Definition of Awareness. 115
3.5.2. Definition of a Person. 115
3.5.3. The Organization of the Human Psyche 118
3.6. Will the Future Belong to the Homo Economicus or to the Homo Informaticus? 120
3.6.1. The current economic model. 120
3.6.2. Gross National Happiness is more important than Gross Domestically Product. 126
3.7. The Anatomy of War 130
3.7.1. War as a phenomenon 130
3.7.2. What are the ingredients of a World War? 134
3.7.2.1 The Prohibition in the US. 134
3.7.2.2. A worldwide Pandemic. 138
3.7.2.3. A Great Recession caused by poor oversight of the markets 139

3.8.2.4. An agricultural depression in an industrialized country. 143
3.7.2.5. A crazy dictator of an industrialized country 144
3.8. Degree of entropy inside the USA following the system theory. 148
3.8.1. Introduction 148
3.8.2. Conspiracy Theories, Social Entropy and Literature. 150
3.8.2. Discussion and Analysis 153
3.8.4. Concluding Remarks 153
Part 4; RELIGION. 155
4.1. Definition of Religion 155
4.2. The Religious Paradigm. 158
4.3. A New View upon Religion. 161
Part 5; Towards a new synthesis. 167
5.1. Introduction 167
5.2. The US as the dominant envelope. 168
5.3 The Upcoming envelopes. 169
5.4. Will Bhutan become the new Switzerland of the Himalaya´s? 170
Appendix A; Mathematical proof for the use of Fournier's Theorem to study periodic phenomena...................................174
Appendix B; Arms races...178
Appendix C; Levitation forces induced by superconductors..179
Appendix D; Technical elaboration upon the proposed mathematical neurological model...181
TABLE I: Best sold literature 1561 - 1862 192
TABLE II: Best sold novels 1863 – 1964 194
TABLE III Best sold novels 1965 - 2015 196
TABLE IV; Best sold non fiction 1917 – 2012.......................197
TABLE V: Dewey Decimal Categories..................................198
TABLE VI; main protagonist types in novels 1863- 1963.....199
TABLE VII; Evolution of the printed book collection of the Library of Congress..201

Part 1; The GEOSPHERE; What we know and don't know about the Cosmos Genesis.

1.1 Einstein's Relativity theory.

Alan Guth believed that the universe bubbled up out of a pre-universal singularity. During a short moment, all the forces and building stones of matter were one. When the Higgs-field symmetries started to break up, followed a hot expansion [3] .

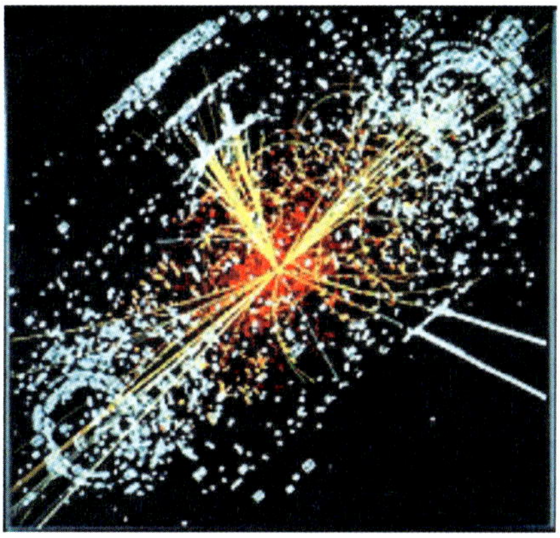

Computerized image of a particle interaction with the Higgs Field. The Higgs Field is an energy field that exists everywhere in the universe. The field is accompanied by a fundamental particle called the Higgs Boson, which the field uses to continuously interact with other particles. As particles pass through the field they are "given" mass, much as an object passing through treacle (or molasses) will become slower. Mass itself is not generated by the Higgs field- the creation of matter or energy would conflict with the laws of conservation. However, mass is "imparted" to particles from the Higgs field, which contains the relative mass in the form of energy. Once the field has endowed a formerly massless particle the particle slows down because it has become heavier. If the Higgs field did not exist, particles would not have the mass required to attract one another, and would float around freely at light speed

Shortly after the Big Bang, the Higgs-field became super cooled and got blocked. This resulted in a false vacuum with latent energy, called false not to confuse it with the empty space. Following insights, already developed by Einstein in his General Relativity Theory[4], such a pseudo-vacuum would cause a repulsive force,

---

[3] Guth, Alan H (1997), The Inflationary Universe, Reading, Massachusetts: Perseus Books, ISBN 0-201-14942-7

resulting in a short period of swelling up, were by the area that would become our visible universe, exploded. When the Higgs-field instead of exploding, coagulated, can it be possible that this created parallel universes[5], divided by energy fields that are dividing the universe into different domains. The swelling up scenario would also explain the whereabouts of the monopoles; they got scattered over the universe during the explosion.

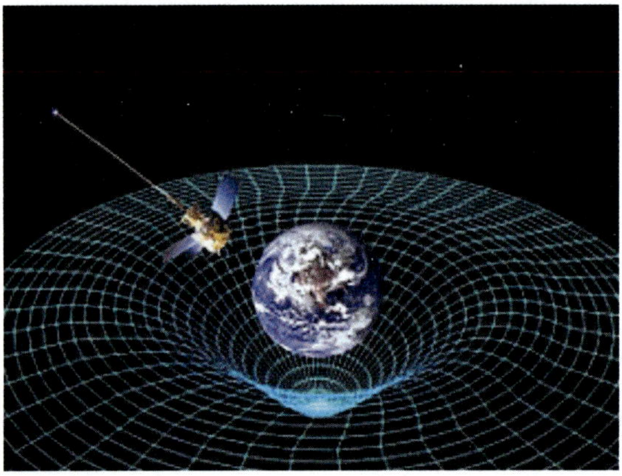

Image of the space-time curvature that causes gravity. The central idea of general relativity is that space and time are two aspects of space-time. Space-time is curved when there is gravity, matter, energy, and momentum. The links between these forces are shown in the Einstein Field Equations. One equation in General relativity is E=mc^2, and there are many more.

If there were things happening before the Big Bang, then they don't have any influence upon what's actually happening. We don't have to take into account their existence because they don't have any consequences for our observations. We can postulate that time started with the Big Bang because previous times are undefined. The Universe is governed by four fundamental forces; Gravity, Electromagnetism, Weak Core energy and Strong Core Energy with the space-time[6] as background.

---

[4] Einstein, A. (1916), "Die Grundlage der allgemeinen Relativitätstheorie", Annalen der Physik 354 (7): 769–822, doi:10.1002/andp.19163540702

[5] Josh Clark (1998-2014). "Do parallel universes really exist?". HowStuffWorks website

[6] Hermann Minkowski, "Raum und Zeit", 80. Versammlung Deutscher Naturforscher (Köln, 1908). Physikalische Zeitschrift 10 104-111 (1909) and Jahresbericht der Deutschen Mathematiker-Vereinigung 18 75-88 (1909). For an English translation, see Lorentz et al. (1952)

Einstein's General Theory of Relativity describes gravitation as a consequence of the curvature of space-time. The gravitational force that two objects exercise upon each other can very nearly be calculated by Newton's universal gravitation law,

$$F = G\frac{m_1 m_2}{r^2},$$

where:

- *F is the force between the masses,*
- *G is the gravitational constant* $(6.673\times10^{-11}\ N\cdot(m/kg)^2)$,
- m_1 *is the first mass,*
- m_2 *is the second mass, and*
- *r is the distance between the centers of the masses.*

Einstein's relativity theory described where gravity is coming from by applying a couple of basic principles and was long time challenged by the Quantum Mechanics Field Theory[7], till the Grand Unified Theory (GUT)[8] created a frame where they could coexist.

1.2. The Quantum Mechanics Field Theory

Quantum Mechanics is the part of physics that can explain why all electronic technology works as it does. Thus QM explains how computers work, because computers are electronic machines. But the designers of the early computer hardware of around 1950 or 1960 did not need to think about QM. The designers of radios and televisions at that time did not think about QM either. However, the design of the more powerful integrated circuits and computer memory technologies of recent years does require QM. The base of QM is the Schrodinger Equation and is up to now the most accurate theory of how subatomic particles behave. It defines something

---

[7] "Quantum Mechanics: The Uncertainty Principle, 1925 - 1927". American Institute of Physics.

[8] Ross, G. (1984). Grand Unified Theories. Westview Press. ISBN 978-0-8053-6968-7.

called the wave function of a particle or system (group of particles) which has a certain value at every point in space for every given time. These values have no physical meaning, yet the wave function contains all information that can be known about a particle or system. This information can be found by mathematically manipulating the wave function to return real values relating to physical properties such as position, momentum (mass times velocity), energy, etc. The wave function can be thought of as a picture of how this particle or system acts with time and describes it as fully as possible.

The wave function can be in a number of different states at once, and so a particle may have many different positions, energies, velocities or other physical property at the same time (i.e. "be in two places at once"). However, when one of these properties is measured it has only one specific value (which cannot be definitely predicted), and the wave function is therefore in just one specific state. This is called wave function collapse and seems to be caused by the act of observation or measurement. The exact cause and interpretation of wave function collapse is still widely debated in the scientific community.

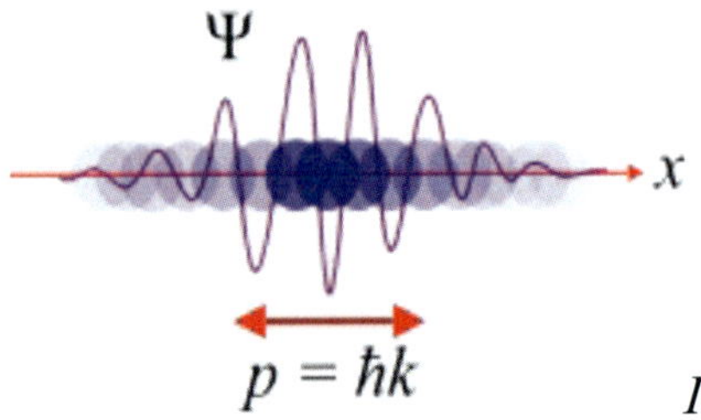

Image of a wave-particle duality

For one particle that only moves in one direction in space, the Schrödinger equation looks like:

$$i\hbar\frac{\partial}{\partial t}\Psi(x,t) = \hat{H}\Psi(x,t)$$

where:

i *is the square root of negative one*

$\hbar$ *is the reduced Planck's constant*

α is the laplacian operator

t is time

x is a place in space

$\Psi(x, t)$ is the wave function

$\hat{H}$ is the Hamiltonian energy operator:

$$\hat{H}\Psi = \hat{T}\Psi + \hat{V}\Psi$$

where:

$\hat{T}$ is the kinetic energy operator. Which is equal to $-\frac{\hbar^2}{2m}(\frac{\partial^2}{\partial x^2}\Psi + \frac{\partial^2}{\partial y^2}\Psi + \frac{\partial^2}{\partial z^2}\Psi)$.

$\hat{V}$ is the potential energy operator equal to $V(x)$, an as yet not chosen function of position.

For one particle that is restricted to a certain region in space (for example: an electron in an atom), the Time-independent Schrödinger Equation can be used, which looks like:

$$\hat{H}\Psi = \mathrm{E}\Psi$$

Where: E is the energy of the particle.

It is also a more general theory of quantum gravity and is an area of active research. It postulates that the gravitational force is mediated by a massless spin-2 particle called the graviton and hypotheses that gravity could have separated from the electronuclear force during the grand unification period.

Gravity is the weakest of the four fundamental forces of the universe and is approximately 10−38 times the strength of the strong force, 10−36 times the strength of the electromagnetic force, and 10−29 times the strength of the weak force. As a result, gravity has a negligible influence on the behavior of sub-atomic particles, and plays no role in determining the internal characteristics of daily matter. Nevertheless, gravity is responsible for causing the stars to form constellations, the planets to orbit the stars, the internal heating up from the stars and for various other phenomena observed throughout the universe. That is because gravity is the only force acting on all particles, has an infinite range, is always attractive and never repulsive and it cannot be absorbed, transformed, or shielded against.

The electromagnetic force is a kind of physical interaction that occurs between electrically charged objects and usually manifests into the form of an electromagnetic field. It plays an important role in determining the internal properties of most common everyday objects. Matter takes its shape as a result of intermolecular forces between individual molecules in matter. Electromagnetic waves mechanics bind particles into orbitals around atomic nuclei to form atoms, which are the building stones of molecules. Processes involved in chemistry arise from interactions between the electrons of neighboring atoms, determined by the interaction between electromagnetic force and the momentum of the electrons.

In traditional electrodynamics, electric fields are described as electric potential and electric current in Ohm's law, magnetic fields are associated with electromagnetic induction and magnetism, while Maxwell's equations describe how electric and magnetic fields are generated and influence each other by charges and currents.

The establishment of the speed of light based on properties of the "medium" of propagation (permeability and permittivity), led to the development of special relativity theorem by Albert Einstein. At high energy the weak force and electromagnetism are unified. During the making of the universe, during the quark epoch, the electroweak force split into the electromagnetic and weak forces.

The weak interaction causes the radioactive decay of subatomic particles and nuclear fission and the theory of it is sometimes referred to as quantum flavor dynamics (QFD), but is best understood in terms of the electro-weak theory (EWT).

The weak interaction is caused by the emission or absorption of W and Z bosons. The known fermions (particles that have half-integer spin) interact because of the weak interaction. Either a fermion is an elementary particle (ex. the electron), or it can be a composite particle (ex. a proton). The weak force has only a short range because the masses of W+, W−, and Z bosons are each far greater than that of protons or neutrons. The force is called weak because its field strength over a given distance is mostly several times of magnitude less than that of other fundamental interactions of nature.

The strong nuclear force is only effective at a distance of a femtometer and ensures the stability of ordinary matter, as it confines the quark elementary particles into hadron particles such as the proton and neutron, the largest components of the mass of ordinary matter. Most of the mass-energy of a common proton or neutron is in the form of the strong force field energy; the individual quarks provide only about 1% of the mass-energy of a proton.

Its function is to bind protons and neutrons together into atoms and is called the nuclear force. It is the residuum of the strong interaction between the quarks that make up the protons and neutrons.

It composes most of the energy that is released during the breakup of a nucleus and is used in nuclear power plants and fission nuclear weapons.

What we call empty space is in fact an enormous background upon which these four forces paint in golf like vibration the Universe.

To understand the Universe, Isaac Newton created a mechanical methodology that still has a great influence on reductionism opinions in contemporary science. Just lately we began to see ourselves as an information process. Since language is the most important tool that we use to transmit information, is it impossible for an individual to give an impartial description that supersedes the limitations of his language. So the universe is something that we invent; into the Jewish culture there goes a saying that states that with every person that dies, a universe gets destroyed.

1.3. Unresolved Issues in the Grand Unified Theory

There is currently no hard evidence that nature is described by a Grand Unified Theory. To conclude this chapter, I want to list the seven most important issues that thwart our understanding of the universe.

1. De limits of my language are the limits of my world. Linguistically research has already established that some languages serve better certain purpose that whatever other language does. You have to see a language as a huge computer program; they all have their specializations. So has linguistically research for example established that the Navajo language would be the best language for discussing nuclear physics, there are 56 different ethnic groups in China but all of them can READ Chinese because it's alphabet is a logographic one (most languages have a phonetically alphabet), the amount and complexity of tunes that a language uses has climatically atavistic roots, etc… It is suggested that the only language that probably could express a comprehensive model of the universe would have to be mathematical. This actual world of what is knowable, in which we are, and which is in us, remains both the material and the limit of our consideration (Schopenhauer[9]).

2. Were Newtonian physics teaches us that the components are important, demonstrated Wiener[10] the importance of the patterns that connects the components to each other. This refers to the Gestalt aspect of a given subject. It comes also to expression into the holistic approach that medical science develops in treating diseases.

3. For the function of our actual model of the Big Bang, the particles of our universe must have been organized with such a perfection that cannot be explained by coincidence. This improbable circumstance is known as the homogametic principle.

---

[9] P Abelsen, Schopenhauer and Buddhism. H Amsterdam, A Schopenhauer – Philosophy East & West, 1993

[10] Norbert Wiener, Cybernetics: Or Control and Communication in the Animal and the Machine. Paris, (Hermann & Cie) & Camb. Mass. (MIT Press) ISBN 978-0-262-73009-9; 2nd revised ed. 1961

4. Another problem is the mystery of the missing magnetic monopoles. Recent research has led to new theories for the existence of magnetic monopoles. In this context scientists are working at a Grand Unified Theory (GUT) who postulates the existence of a magnetic monopole particle. In most quantum field theories the majority of the particles are unstable, but the GUT predicts the existence of a particle called the dyon, whose basic state is a monopole, and who's stable because there is no simpler topological state to decay to. They came to existence as a side effect of the freezing out of the early universe. During research with the superconducting quantum interference device, there have been observations made of monopole events, but the problem lies with the fact that the events cannot be reproduced. Novikov, a Russian astrophysicist, claims that the black holes are entrances to an Einstein-Rosen bridge (a nontraversible wormhole) and could contain magnetic monopoles[11] .

5. There is also the question if the universe will keep expanding, given the fact that the gravitational forces are striving to bring the growth to a standstill. Into the speculative opinion of Alan Guth, the decay of false vacuums (vacuums with the lowest density of energy) at the beginning of the Universe caused the exponential expansion of space.

6. Gravitational effects induced by coupling via superconductivity.

In 1992, a Russian material scientist named Dr. Eugene Podkletnov claimed that he had found an antigravity effect while working with a team of researchers at Tampere University of Technology in Finland. They made a device that caused an anti-gravitational effect by using a ring of superconducting ceramic (Yttrium-barium-copper oxide) spinning at 5000 rpm. An above the rotating dish suspended object showed a variable weight loss from less than 0,5% to better then 2%.

In a follow up to these results the Delta-G experiments were carried by NASA's Marshall Space Flight Center in the USA which attempted to duplicate Podkletnov's claimed anomalous weight loss

---

[11] Igor Dmitriyevich Novikov, Black Holes and the Universe, (translated by Vitaly I. Kisin, Cambridge University Press 1995)

in objects of various compositions suspended above a rotating 12-inch diameter type II ceramic superconductor.

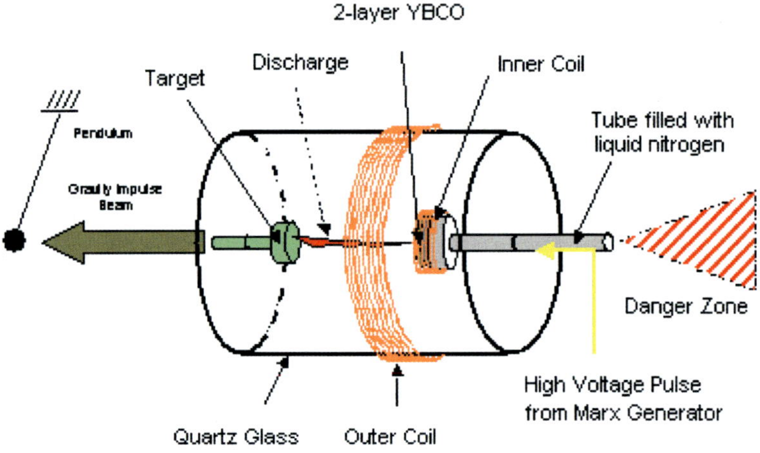

Podkletnov's impulse device

American scientists Douglas Torr and Ning Li[12] at the university of Alabama proposed a theory to explain the anti-gravity effects. This theory can be summarized in a manner parallel to Maxwell's Field equations for electromagnetism for the weak field low velocity limit of general relativity.

Experiments conducted for a variety of different test masses near a rotating type II superconductor were tested by applying an electromagnetic field and a weight loss of up to 5% was reported, which shows that there may exist a coupling between rotating superconductors and gravity. If this result can be more fully understood theoretically, it could lead to the development of a gravitational propulsion system.

---

[12]Li, Ning; Torr, DG (September 1, 1992). "Gravitational effects on the magnetic attenuation of superconductors". Physical Review B46: 5489. Bibcode:1992PhRvB..46.5489L. doi:10.1103/PhysRevB.46.5489. Retrieved March 6, 2015

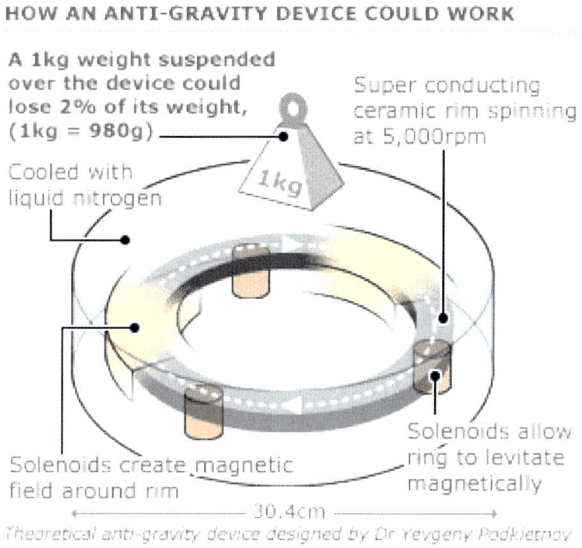
HOW AN ANTI-GRAVITY DEVICE COULD WORK
A 1kg weight suspended over the device could lose 2% of its weight, (1kg = 980g)
Super conducting ceramic rim spinning at 5,000rpm
Cooled with liquid nitrogen
1kg
Solenoids allow ring to levitate magnetically
Solenoids create magnetic field around rim
30.4cm
Theoretical anti-gravity device designed by Dr Yevgeny Podkletnov

A mathematical analysis of the gravitational effects on the magnetic attenuation of superconductors based upon experimentations at the University of Peking can be found in appendix C.[13]

7. Retrocausality.

Although there are many counterintuitive ideas in quantum theory, the idea that influences can travel backwards in time (from the future to the past) is generally not one of them. However, recently some physicists have been looking into this idea, called "retrocausality," because it can potentially resolve some long-standing puzzles in quantum physics. If retrocausality is allowed, then the famous Bell tests can be interpreted as evidence for retrocausality and not for action-at-a-distance—a result that Einstein and others skeptical of that "spooky" property may have appreciated.

One of the main proponents of retrocausality in quantum theory is Huw Price, a philosophy professor at the University of Cambridge. In 2012, Price laid out an argument suggesting that any quantum theory that assumes that

---

[13] L.Liu, Y, Hou, C.Y. He, Z.X. Hao, Effect of magnetization process on levitation force between a superconducting disk and a permanent magnet. Peking University

1) the quantum state is real, and

2) the quantum world is time-symmetric (that physical processes can run forwards and backwards while being described by the same physical laws) must allow for retrocausal influences.

Understandably, however, the idea of retrocausality has not caught on with physicists in general.[14]

[14] Read more at: https://phys.org/news/2017-07-physicists-retrocausal-quantum-theory-future.html#jCp

Part 2; The BIOSPHERE

2.1 Definition of the Biosphere

The biosphere is used to describe the living world ("bio" = life). All of the plants, animals, and other organisms can be found in the biosphere. It is the sum of all ecosystems and biomes on Earth. Thus, the biosphere is the highest level of organization within biology. It is composed of all the other "spheres" of science, such as the ecosphere, atmosphere, lithosphere, and hydrosphere. It reaches to the region of the atmosphere in which birds and insects can still be found. Likewise, the biosphere reaches to the depths of the ocean floor where deep-sea creatures swim or crawl. The "spheres" of the biosphere affect one another just as the organ systems of an organism influence each other. In this way, the biosphere may be viewed as a closed system.

By the most general bio-physiological definition, the biosphere is the global ecological system integrating all living beings and their relationships, including their interaction with the elements of the lithosphere, geosphere, hydrosphere, and atmosphere. The biosphere is postulated to have evolved, beginning with a process of biopoiesis (life created naturally from non-living matter, such as simple organic compounds) or biogenesis (life created from living matter), at least some 3.5 billion years ago[15]. It is an interdisciplinary concept for integrating astronomy, geophysics, meteorology, biogeography, evolution, geology, geochemistry, hydrology and all life and Earth sciences.

---

[15] Campbell, Neil A.; Brad Williamson; Robin J. Heyden (2006). Biology: Exploring Life. Boston, Massachusetts: Pearson Prentice Hall. ISBN 0-13-250882-6.

2.2. Mathematical Model of the Biosphere; Out of the Water Comes all Life.

The following model reflects the system of interconnected ecological, climatic, bioeconomic and socioeconomic processes that are at play in the biosphere.

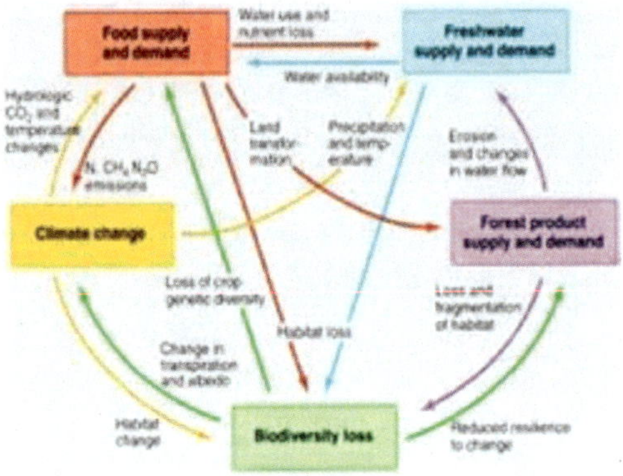

Biosphere dynamics are represented as a diagram of flows energy and matter. The model divides the biosphere into the atmosphere, land and ocean. The atmosphere is characterized by temperature, and by content of carbon dioxide and water vapors, nitrogen and sulfur compounds, and dust. The surface of the land and the ocean is divided into sectors of 4 degrees in latitude and 5 degrees in longitude, within which all processes are considered homogeneous. The vertical heterogeneity of ocean system is considered. [16]

The land surface comprises the smallest areal fraction of the Earth system's major components (e.g., versus atmosphere or ocean with cryosphere). As such, how is it that some of the largest sources of uncertainty in future climate projections are found in the terrestrial biosphere? This uncertainty stems from how the terrestrial biosphere is modeled with respect to the myriad of biogeochemical, physical, and dynamic processes represented (or not) in numerous models that contribute to projections of Earth's future. The model also provides an overview of the processes included in terrestrial biosphere models (TBMs), including various approaches to representing any one given process, as well as the processes that are missing and/or uncertain. We complement this with a comprehensive review of individual TBMs, marking the differences, uniqueness, and recent and planned

---

[16] Mathematical model of global processes in the biosphere. Available from: https://www.researchgate.net/publication/258947366_Mathematical_model_of_global_processes_in_the_biosphere [accessed Jul 24 2018].

developments. To conclude, we summarize the latest results in benchmarking activities, particularly as linked to recent model intercomparing projects, and outline a path forward to reducing uncertainty in the contribution of the terrestrial biosphere to global atmospheric change[17].

A simple but realistic biosphere model has been developed for calculating the transfer of energy, mass and momentum between the atmosphere and the vegetated surface of the earth. The model is designed for use in atmospheric general circulation models.

The vegetation in each terrestrial model grid area is represented by two distinct layers, either or both of which may be present or absent at any given location and time. The upper vegetation layer represents the perennial canopy of trees or shrubs, while the lower layer represents the annual ground cover of grasses and other herbaceous species. The local coverage of each vegetation layer may be fractional or complete but as the individual vegetation elements are considered to be evenly spaced, their root systems are assumed to extend uniformly throughout the entire grid area. Besides the vegetation morphology, the physical and physiological properties of the vegetation layers are also prescribed. These properties determine (i) the reflection, transmission, absorption and emission of direct and diffuse radiation in the visible, near infrared and thermal wavelength intervals; (ii) the interception of rainfall and its evaporation from the leaf surfaces; (iii) the infiltration, drainage and storage of the residual rainfall in the soil; (iv) the control by the photosynthetically active radiation and the soil moisture potential, inter alia, over the stomatal functioning and thereby over the return transfer of the soil moisture to the atmosphere through the root-stem-leaf system of the vegetation; and (v) the aerodynamic transfer of water vapor, sensible heat and momentum from the vegetation and soil to a reference level within the atmospheric boundary layer.

The *Si*mple *B*iosphere (SiB) has seven prognostic physical-state variables: two temperatures (one for the canopy and one for the

---

[17] Joshua B. Fisher, Deborah N. Huntzinger, Christopher R. Schwalm,2 and Stephen Sitch, Modeling the Terrestrial Biosphere, Annual Review of Environment and Resources,Vol. 39:91-123 (Volume publication date October 2014) ,First published online as a Review in Advance on September 15, 2014. https://doi.org/10.1146/annurev-environ-012913-093456

ground cover and soil surface); two interception water stores (one for the canopy and one for the ground cover); and three soil moisture stores (two of which can be reached by the vegetation root systems and one underlying recharge layer into and out of which moisture is transferred only by hydraulic diffusion and gravitational drainage)[18].

Remarks

The model is an attempt to incorporate biophysical realism in a formulation appropriate to existing atmospheric general circulation models. All the current elements of the SiB model are necessary for a realistic calculation of the radiation, heat and momentum fluxes R_n, λE, H and T.

In addition, equations must be added that will correctly partition the GCM produced grid area average convective rainfall into grid average interception storage, interception loss, surface runoff and soil infiltration terms. To treat all of the GCM produced rainfall as it were large-scale, and therefore uniformly distributed over the grid area, may result in significant error.

---

[18] P. J. Sellers and Y. Mintz, A Simple Biosphere Model (SIB) for Use within General Circulation Models, https://doi.org/10.1175/1520-0469(1986)043<0505:ASBMFU>2.0.CO;2

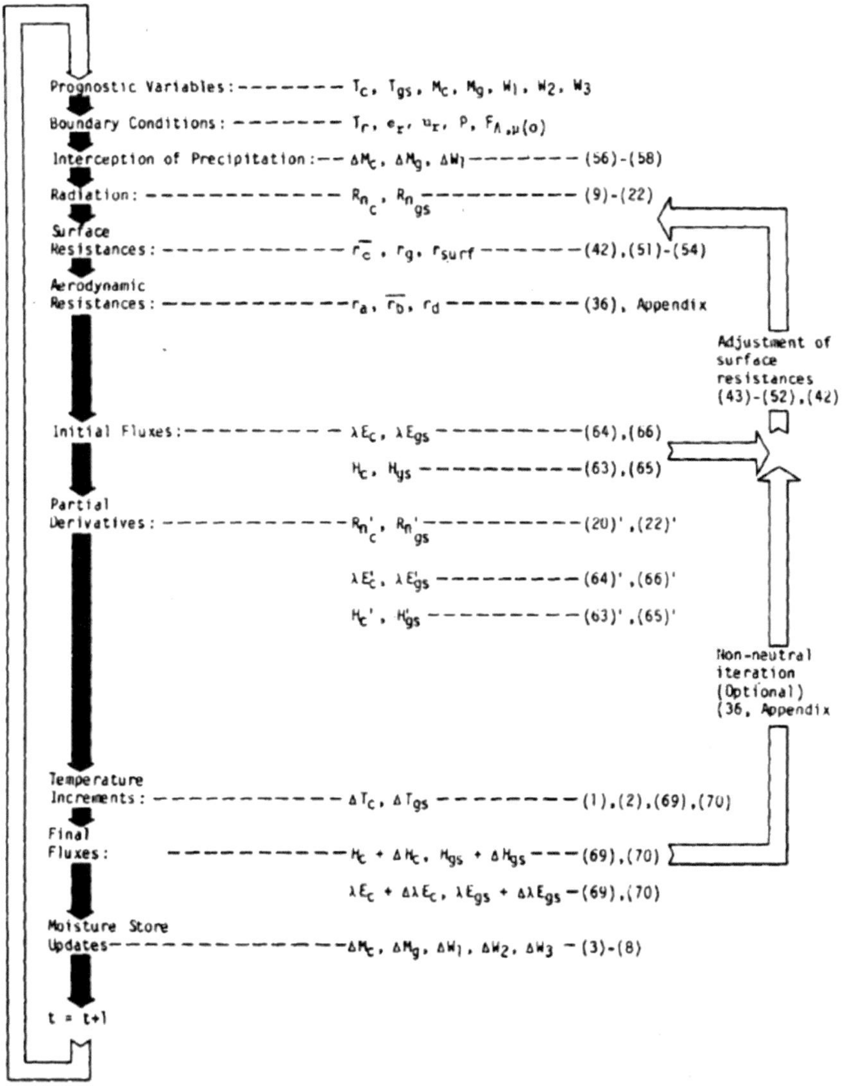

Figure 5: Sequence of calculations used to advance the prognostic variables of the SiB by one-time step. Numbers in parentheses refer to equations in the model. Dash symbol refers to partial differentiation with respect to T_c and T_{gs}

2.3. The Evolution of Life.

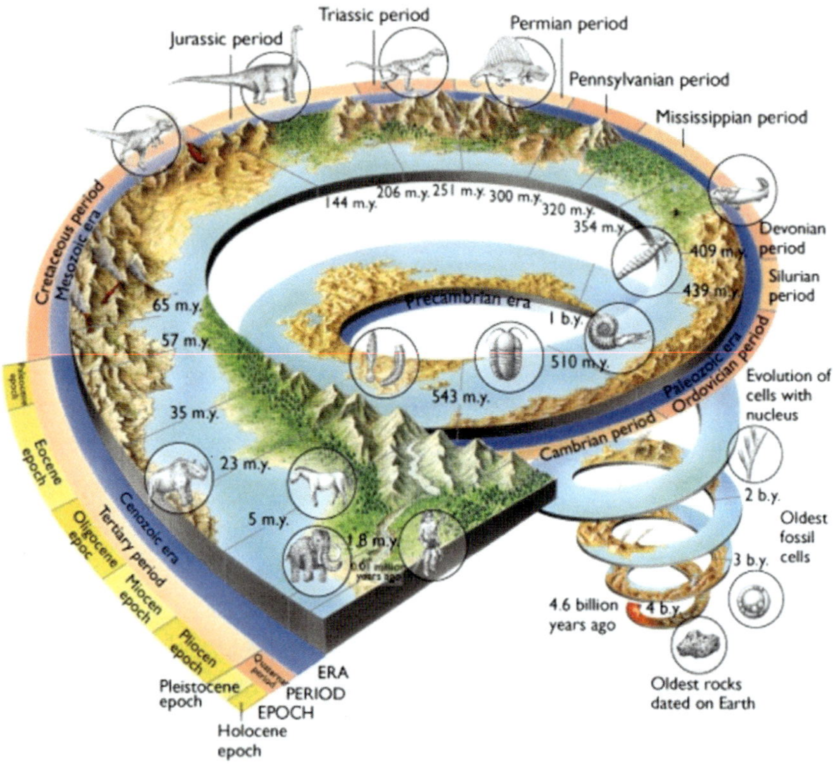

Evolutionary biology has long required this approach, due in part to the complexity of population-level processes and to the long time scales over which evolutionary processes occur. Indeed, the “modern evolutionary synthesis” of the 1930s and 40s—a pivotal moment of intellectual convergence that first reconciled Mendelian genetics and gene frequency change with natural selection—hinged on elegant mathematical work by RA Fisher, Sewall Wright, and JBS Haldane. Formal (i.e., mathematical) evolutionary theory has continued to mature; models can now describe how evolutionary change is shaped by genome-scale properties such as linkage and epistasis[19], complex demographic variability, environmental variability, and individual and social behavior, within and between species[20].

---

[19] Epistasis is the phenomenon where the effect of one gene (locus) is dependent on the presence of one or more 'modifier genes', i.e. the genetic background.

Darwin's theory of natural selection represents one such model, and many others have followed since; for example, Muller proposed that genetic recombination might evolve to prevent the buildup of deleterious mutations ("Muller's ratchet"), and the "Red Queen hypothesis" proposes that co-evolution between antagonistically interacting species can proceed without either species achieving a long-term increase in fitness. A clear verbal model lays out explicitly which biological factors and processes it is (and is not) considering and follows a chain of logic from these initial assumptions to conclusions about how these factors interact to produce biological patterns.

However, evolutionary processes and the resulting patterns are often complex, and there is much room for error and oversight in verbal chains of logic. In fact, verbal models often derive their influence by functioning as lightning rods for debate about exactly which biological factors and processes are (or should be) under consideration and how they will interact over time. At this stage, a mathematical framing of the verbal model becomes invaluable. It is this proof-of-concept modeling on which we focus below.

A century after Darwin published his comprehensive treatment of sexual reproduction, John Maynard Smith used a simple mathematical formalization to identify a biological paradox: why is sexual reproduction ubiquitous, given that asexual organisms can reproduce at a higher rate than sexual ones by not producing males (the "2-fold cost of sex")? Increased genetic variation resulting from sexual reproduction is widely thought to counteract this cost, but simple proof-of-concept models quickly revealed both a flaw in this verbal logic and an unexpected outcome: sex need not increase variation, and even when it does, the increased variation need not increase fitness. Subsequent theoretical work has illuminated many factors that facilitate the evolution and maintenance of sex. Otto and Nuismer, for example, used a population genetic model to examine the effects on the evolution of sex of antagonistic interactions

---

[20] Maria R. Servedio , Yaniv Brandvain, Sumit Dhole, Courtney L. Fitzpatrick, Emma E. Goldberg, Caitlin A. Stern, Jeremy Van Cleve, D. Justin Yeh, Not Just a Theory—The Utility of Mathematical Models in Evolutionary Biology, Published: December 9, 2014 https://doi.org/10.1371/journal.pbio.1002017

between species. Such interactions were long thought to facilitate the evolution of sex. They found, however, that these interactions only select for sex under particular circumstances that are probably relatively rare. Although these predictions might be difficult to test empirically, their implications are important for our conceptual understanding of the evolution of sex.

2.4. The Human Brain.

The human brain is the apex of the terrestrial evolution and its study, neuroscience, is a multidisciplinary branch of biology, which combines physiology, anatomy, molecular biology, developmental biology, cytology, mathematical modeling and psychology to understand the fundamental and emergent properties of neurons and neural circuits. The understanding of the biological basis of learning, memory, behavior, perception and consciousness has been described by Eric Kandel[21] as the "ultimate challenge" of the biological sciences.

2.4.1. The Neurophilosophy of Sperry.

The thinking of Sperry[22] is a good temporary guide to the goals, possibilities and values that are leading towards a better understanding of the forces that control the universe and created humanity. His split-brain-studies were leading him towards his neuro-philosophical theory, whereby de mind gets attributed with causal, leading functions. Sperry observed subtle behavioral changes that occurred by cutting the brain-beam, indicating the way the brain is organized as a substrate of the mind.

The studies demonstrated that the left and right hemispheres are specialized in different tasks. The left side of the brain is normally specialized in taking care of the analytical and verbal tasks. The left side speaks much better than the right side, while the right half takes care of the space perception tasks and music, for example. The right hemisphere is involved when you are making a map or giving directions on how to get to your home from the bus station. The right hemisphere can only produce rudimentary words and phrases, but contributes emotional context to language. Without the help from the

---

[21] Kandel, Eric R. (2012). *Principles of Neural Science*, Fifth Edition. McGraw-Hill Education. p. 5. ISBN 978-0071390118. The last frontier of the biological sciences – their ultimate challenge – is to understand the biological basis of consciousness and the mental processes by which we perceive, act, learn, and remember.

[22] "*Science and moral priority: merging mind, brain and human values.*" Convergence, Vol. 4 (Ser. ed. Ruth Anshen) New York: Columbia University Press (1982)

right hemisphere, you would be able to read the word "pig" for instance, but you wouldn't be able to imagine what it is.[23]

Basing himself upon his research, Sperry argues that the mind is budding out of the matter; it is not only rising up out of it but also supersedes the matter to, secondarily, take control over it.

Out of his extensive research it appears that both hemispheres contain their own central processing unit, were by through own cognitive- and affective possibilities a consciousness is constructed as a global entity

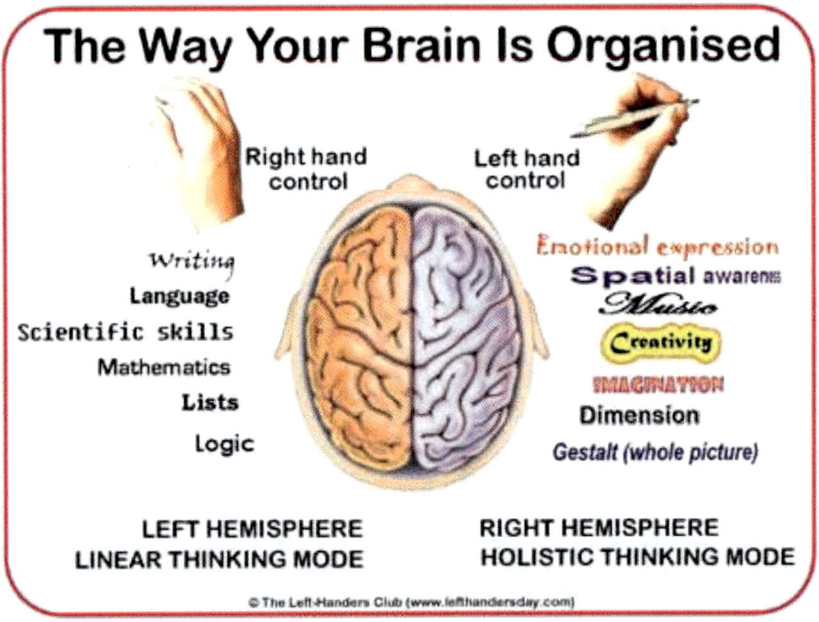

When the unity of the brain is breached, it is observed that both halves are carrying consciousness, but also contribute both to the mental condition of a person as a resultant of the whole working of the brain, generating an undividable union. With the resulted, brain-steered mental data a holistically image of reality was created that was controlling behavior.

The neuro-philosophical implications can be resumed as following;

---

23 The Split Brain Experiments". Nobel Media. Retrieved 17/02/2015

1. The human brain has such a degree of functional complexity and configurationally organization that the mind arises from it as a pattern characteristic.
2. This mind exercises next a downwards regulation upon the brain.
3. This stems with the hierarchy of causal forces in the universe; The components of a subsystem at every level are building at the characteristics of a system, were by the system develops its own functioning in time and space and determines the relational characteristics of the components from what it is constructed

2.4.2. The Mathematical Neurological Model of Johannes Friedrich and Máté Lengye

There exist a multitude of neurological mathematical models, but the one developed by Johannes Friedrich and Máté Lengyel [24], two researchers from the University of Cambridge, is the first biologically realistic account of the process, and can predict not only behavior, but also neural activity. I must admit that every effort to follow their discourse will most likely put a strain on one's own neurobiological processes.

They presented their findings as a spiking neural network. A major advantage of formal spiking neurons such as the Spike Response Model is their simplicity which has several important consequences.

- It is possible to simulate neural networks with many neurons at a reasonable numerical cost.
- Network properties such as the mean firing rate of neurons in a network of connected spiking units can be studied analytically using tools from mathematical probability theory, statistical physics, and bifurcation theory.
- Questions of neural coding can be addressed in a transparent fashion.
- Simple spiking neuron models can be fitted to experimental data.
- Compared to standard leaky integrate-and-fire models the Spike Response Model allows to cover refractoriness (the amount of time it takes for an excitable membrane to be ready for a second stimulus once it returns to its resting state following an excitation).

The network provably solves the difficult online value estimation problem underlying goal-directed decision making in a near-optimal way and reproduces behavioral as well as neuro--

---

[24] Johannes Friedrich and Máté Lengyel, Goal-Directed Decision Making with Spiking Neurons, Journal of Neuroscience 3 February 2016, 36 (5) 1529-1546; DOI: https://doi.org/10.1523/JNEUROSCI.2854-15.2016

physiological experimental data on tasks ranging from simple binary choice to sequential decision making.

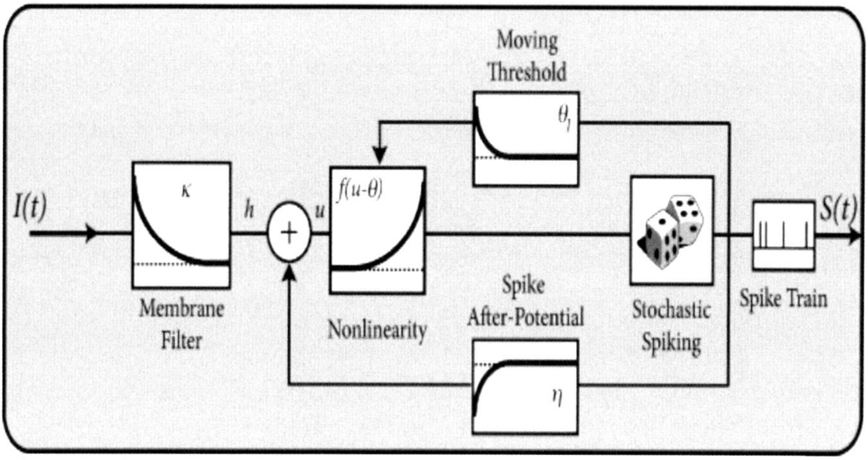

The membrane potential in the spike response model is given by

$$u(t)= \eta\ (t\text{-}t) + {}_{0}\int^{\infty} K(t\text{-}t_{'1}s)\ I(t-s)ds$$

- here t\` is the firing time of the last spike
- η describes the form of the action potential
- K the linear response to an input pulse
- I(t) is a stimulating current

The next spike occurs if the membrane potential u hits a threshold $\theta_{(t-t')}$ in which case t′ is updated.

Their model uses local plasticity rules, which is the ability of synapses to strengthen or weaken over time, in response to increases or decreases in their activity. Since memories are postulated to be represented by vastly interconnected neural circuits in the brain, synaptic plasticity is one of the important neurochemical foundations of learning and memory.

The plasticity rules of the model describe the synaptic weights of a simple neural network to achieve optimal performance and solves one-step decision-making tasks, commonly considered in neuro-economics, as well as more challenging sequential decision-making tasks within 1s. These decision times, and their parametric dependence on task parameters, as well as the final choice probabilities match behavioral data, whereas the evolution of neural activities in the network closely mimics neural responses recorded in frontal cortices during the execution of such tasks.

Their theory provides a principled framework to understand the neural underpinning of goal-directed decision making and makes novel predictions for sequential decision-making tasks with multiple rewards.

Goal-directed actions requiring prospective planning pervade decision making, but their circuit-level mechanisms remain elusive. They show how a model circuit of biologically realistic spiking neurons can solve this computationally challenging problem in a novel way. The synaptic weights of their network can learn by using local plasticity rules such that its dynamics devise a near-optimal plan of action.

By systematically comparing their model results to experimental data, they show that it reproduces behavioral decision times and choice probabilities as well as neural responses in a rich set of tasks. The results thus offer the first biologically realistic account for complex goal-directed decision making at a computational, algorithmic, and implementational level.

Research in animal learning and behavioral neuroscience has given rise to the view that reward-based decision making is governed by (at least) two distinct strategies: a habit system, which relies on cached associations between actions or situations and their long-run future values, and a goal-directed system, which involves prospective planning and comparison of action outcomes based on an internal model of the environment. These systems have their formal counterparts in theories of model-free and model-based reinforcement learning, respectively.

Temporal-difference algorithms of model-free learning account for both behavioral and neuro-imaging data regarding habit-based decision making. Similarly, model-based reinforcement learning algorithms have provided a powerful framework to account for goal-directed behavior and to identify some of the key brain areas involved in it. However, whereas the reward prediction error theory of the phasic responses of dopaminergic neurons has enjoyed prominent success in providing a circuit-level description of the implementation of model-free decision making, much less is known about the neural circuit mechanisms of model-based decision making.

Their work proposes the first instantiation of model-based reinforcement learning in a biologically realistic network of spiking neurons and shows how such a network can compute a quantity that forms the basis of model-based decision making: the value of the best action in a given situation. This optimal value expresses the prediction of cumulative future reward after the execution of that action according to an internal model of how actions lead to future situations and rewards, and if all later actions will also be chosen optimally so as to maximize the cumulative future reward.

Thus, there is a strongly nonlinear, recursive relationship between the optimal values of different actions. This is formalized by the so-called Bellman optimality equation, a necessary condition for optimality associated with the mathematical optimization method known as dynamic programming.

2.4.3. The Bellman optimization equation in our decision making.

4.2.3.1. An illustration of the working of the algorithm

Allow me to illustrate the working of the Bellman equation in our decision making with a little fairytale.[25]

Once upon a time there was a little girl who got a cake. The girl decided to eat the cake all alone. But she was not sure when she wanted to eat the cake. First, she thought of eating the whole cake right away. But then, nothing would be left for tomorrow and the day after tomorrow.

---

[25] Kurt Schmidheiny/Manuel Wlti Doing Economics with the Computer, Session 5: The Cake-Eating Problem, Universität Bern, Summer Term 2002

Well, on the one hand, eating cake today is better than eating eat it tomorrow. On the other hand, eating too much at the same time might not be the best. She imagined that the first mouthful of cake is a real treat, the second is great, the third is also nice. But the more you eat, the less you enjoy it. In the end you're almost indifferent, she thought. So, she decided to eat only a bit of the cake everyday. Then, she could eat everyday another first mouthful of cake. The girl knew that the cake would be spoiled if she kept it more than nine days. Therefore, she would eat the cake in the first ten days. Yet, how much should she eat every day? She thought of eating everyday a piece of the same size. But if eating cake today is better than waiting for tomorrow, how can it possibly be the best to do the same today as tomorrow? If I ate just a little bit less tomorrow and a little bit more today, I would be better off, she concluded.

— And she would eat everyday a bit less than the previous day and the cake would last ten days long and nothing would be left in the end.

4.2.3.2. A formal analysis of the story.

Let us solve the girl's problem formally. Assume that the preferences for cake are in any period described by the following utility function:

$$U(c_t) = lnc_t \qquad (1)$$

where, c_t is the amount of cake consumed in period t. This utility function yields decreasing marginal utility. Future consumption is discounted with the time-preference factor $\beta < 1$. The present value in period 0 of the whole consumption path $(c0, c1, ..., cT)$ is, therefore,

$$V(c\,, c\,, ..., c\,) = \sum_{i=0}^{T} \beta^t U(c_t) \qquad (2)$$

where T is the last day of consumption. In the above story T is 9 as "today" is 0. These preferences are called intertemporally additive. The person tries to maximize (2) by choosing the consumption in period $t = 0, ..., T$. The cake size in period t is the previous size less the previous consumption:

$$k_t = k_{t-1} - c_{t-1} . \quad (3)$$

Consumption in any period must not exceed the cake size in that period. Therefore, the cake size must be nonnegative in any period:

$$k_t \geq 0 \text{ for } t = 1 \ldots T+1. \quad (4)$$

4.2.3.3. The Relevance of the Story

The cake-eating problem yields the basic mathematical structure of the optimal growth models in modern macroeconomics. The famous Ramsey Model can in principle be reduced to the above-mentioned problem. A society must choose between consumption or investments. The more a society invests, the less it can consume instantaneously but the more it can produce, and hence consume, in the future. Models of optimal growth determine the optimal level of investment in any period.
The cake-eating´s problem can be solved analytically. The optimal consumption path satisfies the so-called *Euler equation*:

$$U'(c_t) = \beta U'(c_{t+1}) \quad (5)$$

The Euler equation has an intuitive interpretation: at a utility maximum, the consumer cannot gain from feasible shifts of consumption between periods. A one-unit reduction in period t consumption lowers U_t by U_t'. This unit saved can be shifted to period $t + 1$ where it raises utility by U'_{t+1}. Discounted to period t this is $\beta \cdot U'_{t+1}$. In the optimum, these two quantities must be equal. Using utility function (1), the Euler-equation (5) is

$$\beta \cdot U'_{t+1}. \quad (6)$$

In the optimum nothing is left in the end

$$k_{T+1} = 0. \quad (7)$$

Recursive insertion of (3) and (6) in (7) yields the optimal initial period consumption: Recursive insertion of (3) and (6) in (7) yields the optimal initial period consumption:

$$c_0 = \frac{1-\beta}{1-\beta^{T+1}} k_0 . \tag{8}$$

Together with (6) the optimal consumption path is exactly described.

2.4.3.4. Method

The Bellman process breaks a dynamic optimization problem into a sequence of simpler sub-problems, as Bellman's "principle of optimality" prescribes, which makes the computation of optimal values challenging. Indeed, goal-directed decision making is typically considered as flexible but slow, in contrast to inflexible but fast habitual decision making and a Bellman equation is a recursion for expected rewards.

Goal-directed planning is performed in this network on the time scale of hundreds of milliseconds in simpler one-step neuroeconomic, and even more complex sequential, decision-making tasks by using neurally plausible network dynamics. It is demonstrated that the network indeed competently solves the Bellman equation by establishing its performance on an example artificial decision-making task (after benchmarking it more extensively on several standard tasks from the reinforcement learning literature). Results are presented from a set of simulations illustrating the success of the model at accounting for a variety of experimental findings involving goal-directed action selection.

This model reproduces behavioral and neurophysiologic data on tasks ranging from simple binary choice to multistep sequential decision making. It also makes predictions that are testable at the level of behavior and in neural responses for a new, multiple-reward variant of an already existing sequential decision-making task.

I have limited the technical discussion of this model to a brief description of the neurobiological processes that they consider. A more detailed technical elaboration of the above proposed model can be found in appendix D. The mathematization of the used optimal values, the neural network dynamics and their relation to other programming algorithms have been omitted. The assertion that this information is available seemed to cover the needs of this essay, and I gladly refer the curious reader who would like to find more about it to the research paper of Friedrich and Lengyel.

2.4.4. Analysis and Discussion.

Although the above described model is a big leap forward into our understanding of the functioning of the brain, still many issues remain unsolved. These problems include:

Consciousness: What is the neural basis of subjective experience, cognition, wakefulness, alertness, arousal, and attention? Is there a "hard problem of consciousness"? If so, how is it solved? What, if any, is the function of consciousness?

Sensation/Perception; How does the brain transfer sensory information into coherent, private percepts? What are the rules by which perception is organized? What are the features/objects that constitute our perceptual experience of internal and external events? How are the senses integrated? What is the relationship between subjective experience and the environment?

From childhood, we are taught that the human body has five senses. I'm sure most readers can recite them: sight, hearing, touch, taste, and smell. This list has remained unchanged since the time of Aristotle. To most people, a "sixth sense" refers either to one outside the realm of the scientific, or one that simply does not exist in most humans.

Biologists working on the electrical and magnetic senses of animals have already claimed the sixth sense. Some species of eels, for example, generate electrical fields around themselves through which they sense objects in their environment, even in the dark. Sharks and rays detect, with astonishing sensitivity, the body electricity of potential prey. Various species of migratory fish and birds have a magnetic sense, a biological compass that enables them to respond to Earth's magnetic field. Since the human genome evolved from those species, it can be reasonably argued that this sense is present in latent form in our sensory nervous system but has been suppressed by our perception in favor of the five more commonly used senses. We all know people who have an uncanny ability to orientate themselves in a strange environment. The unexplained capacity of the augurer to find water layers who are hidden under meters of soil or rocks may well be related to an ability of that person to detect changes in the earth's magnetic field where water layers occur.

The seventh sense is related with a certain perception of the current of time. There exist people who have a cunning awareness of the hour of the day without having to look upon a clock or the use of other external indicators that help other people to estimate the passage of time (the position of the sun being the most obvious one). It is also often connected with intuitive behavior & premonition. Retro-causal phenomena and the nature of time are intensively discussed themes among natural physicians.

"The history of the senses in science and culture suggests that parsing the senses introspectively allows for both greater and fewer than five senses. The physiological models suggest incredible complexity – and many more than five systems, while the kinds of media we detect allow for much less. Given all the advances we have made towards their understanding, perhaps it is time to abandon the notion that we have five senses and seek a more scientifically valid conception"[26]

Noogenesis - the emergence and evolution of intelligence: What are the laws and mechanisms - of new idea emergence (insight, creativity synthesis, intuition, decision-making, eureka); development (evolution) of an individual mind in the ontogenesis, etc.?

2.4.5. Conclusion

Notwithstanding the many unresolved issues that remain in neuroscience, I retain for the purpose of this essay that the human brain proceeds in two different ways.

1. Atavistic, instinctive, and routinely behavior is governed by cached neural plasticity rules.

[26] Gordon, M.S., Looking Back: Finding the Senses, The Psychologist Vol. 25 p.908 - 909, Dec. 2012

2. The long termed and/or goal directed behavior is following a process that can be described by a Bellman optimization equation.

This all happens at a speed determined by the membrane spike potential of the involved neurons.

I concur with Sperry´s postulation that the mind knows a material preconditioning but has also a will. This free will supersedes its material origins but stays anchored into the stream of the creation and its laws.

Part 3. The Noosphere

The Noosphere can be viewed as a vast network of computations wherein information is created, transformed, and destroyed. This information often exhibits patterns, or statistical regularities that can be expressed mathematically.

4.2. The Noosphere and Memetics.

The word —meme is a neologism coined by Richard Dawkins in The Selfish Gene (1976) and defined as a self-reproducing and propagating information structure analogous to a gene in biology. Dawkins focused on the meme as a replicator, analogous to the gene, able to affect human evolution through the evolutionary algorithm of variation, replication, and differential fitness.

A meme is information which propagates, persists, and has impact. To distinguish memes from other kinds of information, an elaboration of the definition invokes a threshold for propagation and persistence and employs Shannon 's definition of information as that which reduces uncertainty. Impact might take place in the individual, in terms of neurophysiologic changes (e.g., as detected by a brain scan) or manifestations of behavior; or it might take place in society (groups of individuals) as manifested by behavior.

As a practical approach, the metrics, initially defined for evaluating memes, as shown in Figure 1, include: propagation, persistence, impact, and entropy.

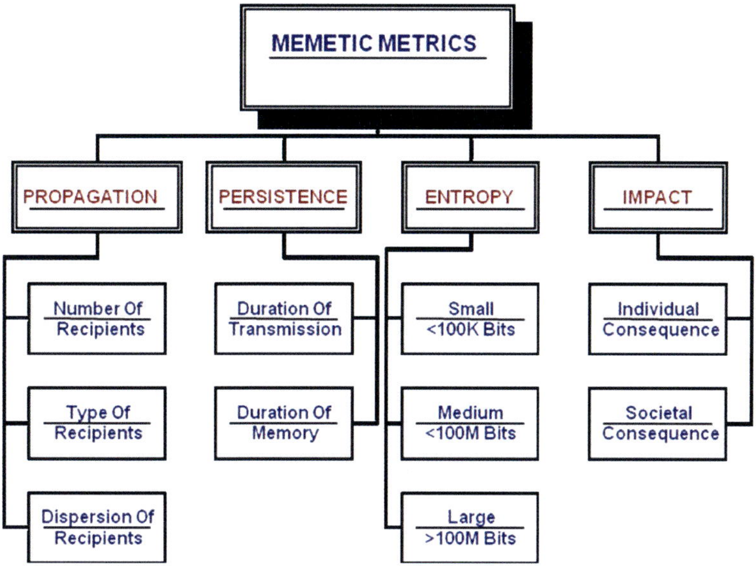

Figure 1: Memetic Metrics and Submetrics

The submetrics for the propagation metric include the number, type, and dispersion of recipients of the meme. Depending on the problem under consideration, the type of recipients might be characterized or categorized by their economic, social, or educational class, ethnicity or culture, religion, gender, age, tribe, politics, etc., while the dispersion of recipients might categorized as local, tribal, familial, regional, national, global, etc. The submetrics for the persistence metric distinguish between the duration of transmission of the meme and the duration of the meme in memory or storage. The submetrics for the entropy metric distinguish among small, medium, and large memes, which (using an order of magnitude rule) are characterized as less than or equal to 100K bits, less than or equal to 100M bits, and greater than 100M bits. Submetrics for the impact metric distinguish between the impact (or potential impact) of the meme on the individual (i.e., individual consequence) and its impact (or potential impact) on society as a whole (i.e., societal consequence). For the initial exercises, a multivariate decision tool was used– the Analytic Hierarchy Process (AHP) – to weight the metrics and submetrics.

Meme Transmission As shown in Figure 2 (which replicates Claude Shannon ‘s iconic schematic of a general communications system), a meme is transmitted after either being created in the mind of an individual or re-transmitted after being received by an individual from elsewhere. Arriving at a new potential host, the meme is received and decoded.

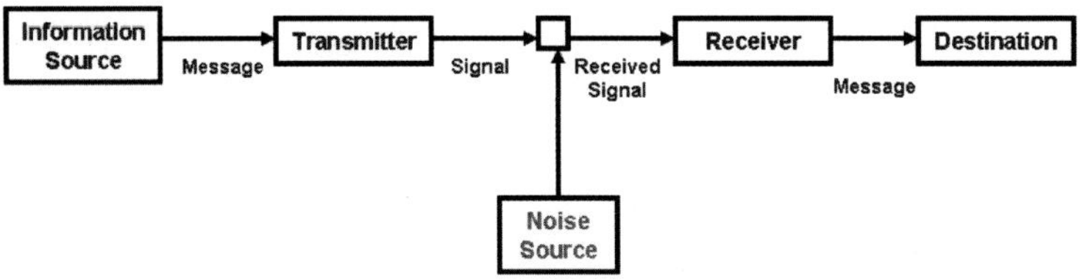

Figure 2: Meme Transmission Replicates Claude Shannon’s Iconic Schematic of a General Communications System

Culture can be defined as the total pattern of behavior (and its products) of a population of agents, embodied in thought, action and artifacts, and dependent upon the capacity for learning and transmitting knowledge to succeeding generations.

Although people often make the decision to spread a meme or proceed unconsciously, this process is influenced by the meme. Some memes are viewed as important, and hence spread to others after a conscious and sometimes rational evaluation; some memes exploit aspects of cognition or emotion to bias their hosts to spread them. Natural selection favors memes that are good at reproducing, which suggests that in time there will exist many memes that are very efficient replicators. Their accuracy is irrelevant for their survival, only their ability to replicate and find new hosts; memes that interest people and encourage them to spread the meme will thrive at the expense of less attractive versions.

Memes can become extinct for the same reasons as parasites. Their biological dynamic exclusively concerns their own replication - in this respect they are strictly "selfish" to use the metaphor that Dawkins applied to genes - and consequently are neither detrimental nor beneficial to their hosts by necessity.

3.2. Evolution and Human History.

3.2.1. The Mathematical Model of Claudio Maccone.

The mathematical model developed by Claudio Maccone[27] is capable of merging Darwinian Evolution and Human History[28] into a single mathematical scheme and was coming to the following conclusions:

1. Darwinian Evolution over the last 3.5 billion years is defined as one particular realization of a certain stochastic process called Geometric Brownian Motion (GBM).

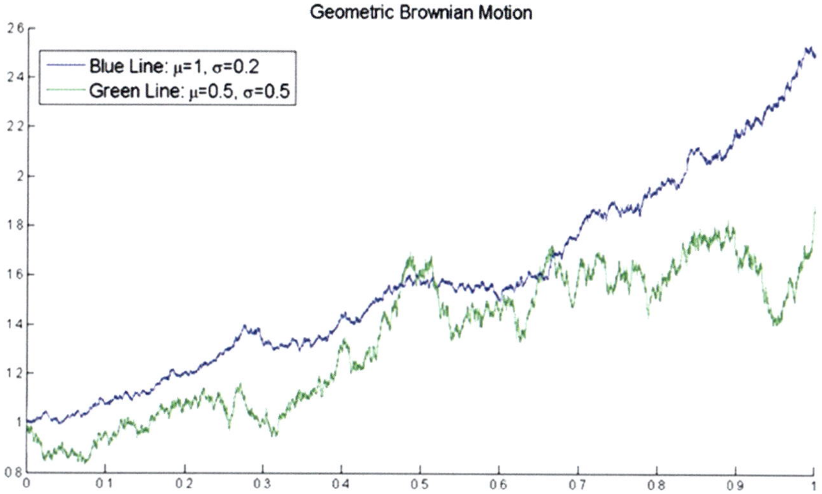

GBM. Two particular realizations of the stochastic process called Geometric Brownian Motion (GBM) taken from the Wikipedia site http://en.wikipedia.org/wiki/Geometric_Brownian_motion. Their mean values are the exponential with different values of A and B for each shown stochastic process.

2. This GBM yields the fluctuations in time of the number of species living on Earth. Its mean value curve is an increasing

---

[27] Claudio Maccone, SETI, Evolution and Human History Merged into a Mathematical Model, Cambridge University Press, 2013.
https://pdfs.semanticscholar.org/4215/2e4bc9f0c6c15efd107fa1223485f58a9c4d.pdf

[28] SETI is a scientific experiment, based at UC Berkeley, that uses Internet-connected computers in the Search for Extraterrestrial Intelligence

exponential curve, i.e. the exponential growth of Evolution. Geometric Brownian motion is used to model stock prices in the Black–Scholes model and is the most widely used model of stock price behavior. A GBM process only assumes positive values, just like real stock prices.

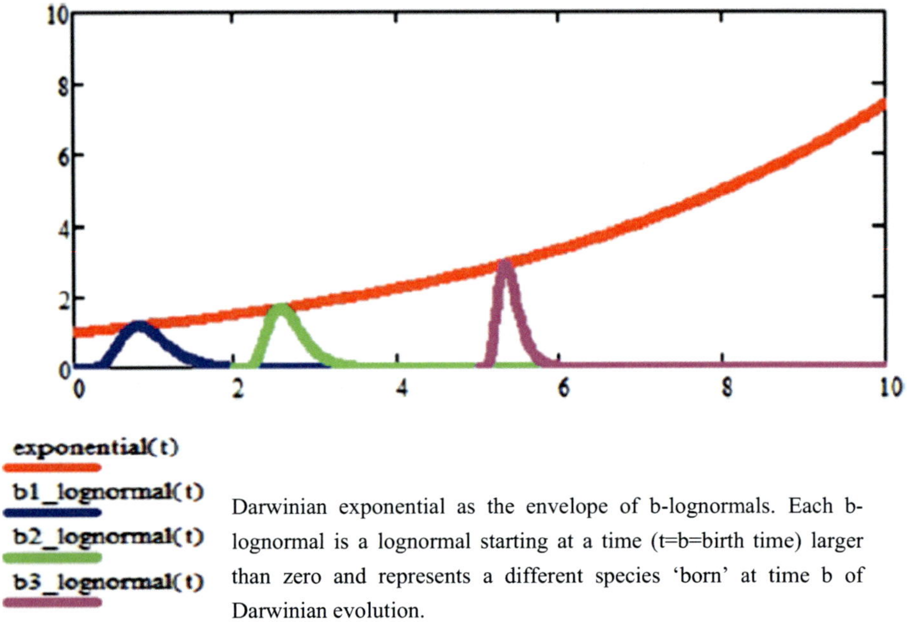

Darwinian exponential as the envelope of b-lognormals. Each b-lognormal is a lognormal starting at a time (t=b=birth time) larger than zero and represents a different species 'born' at time b of Darwinian evolution.

3. He calls "b-lognormals" those lognormals starting at any positive time b ("birth") larger than zero. Then the exponential growth curve becomes the geometric locus of the peaks of a one-parameter family of b-lognormals: this is his way to re-define cladistics. He used the GBM to calculate the increase of numbers of species and as a key to stochastic evolutions of all kinds while admitting to the fact that extinctions (or vertical downfalls) happen from time to time. Cladistics is the science describing when new forms of life developed during Evolution. Cladistics is thus the science of lineages, i.e. phylogenetic trees, like the one shown for instance in next figure, and it is today strongly based on computer codes, in turn based on high-level mathematics.

Example of a horizontal cladogram with the not named ancestor to the left.

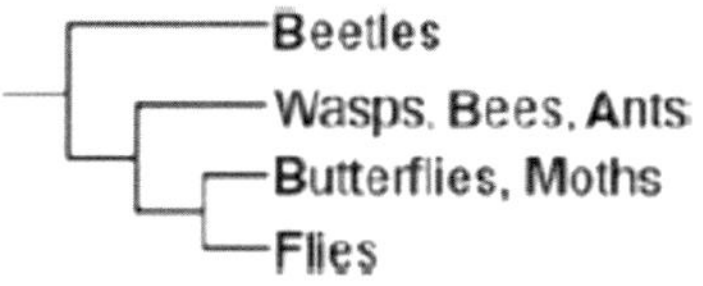

His innovative contribution to cladistics and cladograms, is to put the horizontal axis of time below them, and then realize that the cladograms branches are exponential functions of the time. In other words, these exponential arches are either increasing in time, or decreasing, or just staying constants (i.e. they are just horizontal lines, like the ones in previous figure), but the length of these exponential arches is as long as the species they represent survived during the course of evolution.

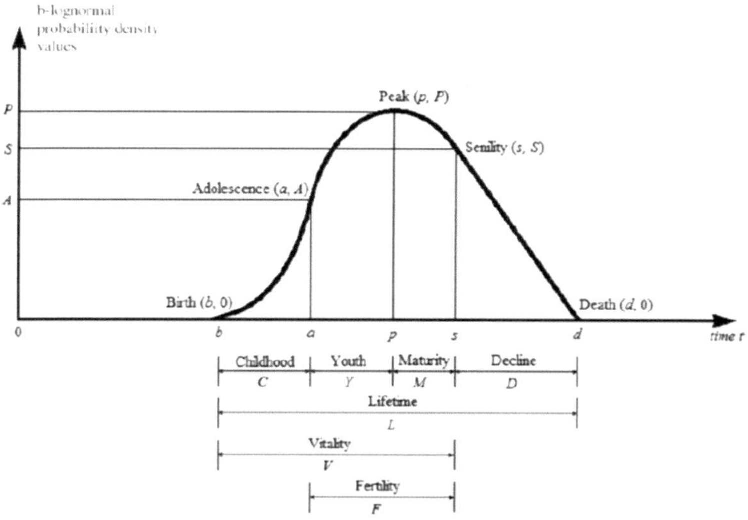

Lifetime of all living beings, i.e. finite b-lognormal: definitions of the basic instants of birth (b= starting point on the time axis), adolescence (a=ascending inflexion abscissa, with ordinate A), peak (p=maximum point abscissa, with ordinate P), senility (s=descending inflexion abscissa, with ordinate S) and death (d=death abscissa=intercept between the time axis and the straight line tangent to the b-lognormal at the descending inflexion point). Also defined are the obvious single-time-step-spanning segments called childhood (C= a−b), youth (Y=p−a), maturity. (M=s−p), decline (D= d−s). In addition, also defined are the multiple-time-step-spanning segments of the all-covering lifetime (L=d−b), vitality (V=s−b) (i.e. lifetime minus decline) and fertility (F=s−a) (i.e. adolescence to senility).

4. b-lognormals may be also be interpreted as the lifespan of any living being (a cell, or an animal, a plant, a human, or even the historic lifetime of any civilization). Applying this new mathematical apparatus to Human History, leads to the discovery of the exponential progress between Ancient Greece and the current USA as the envelope of all b-lognormals of Western Civilizations over a period of 2500 years.

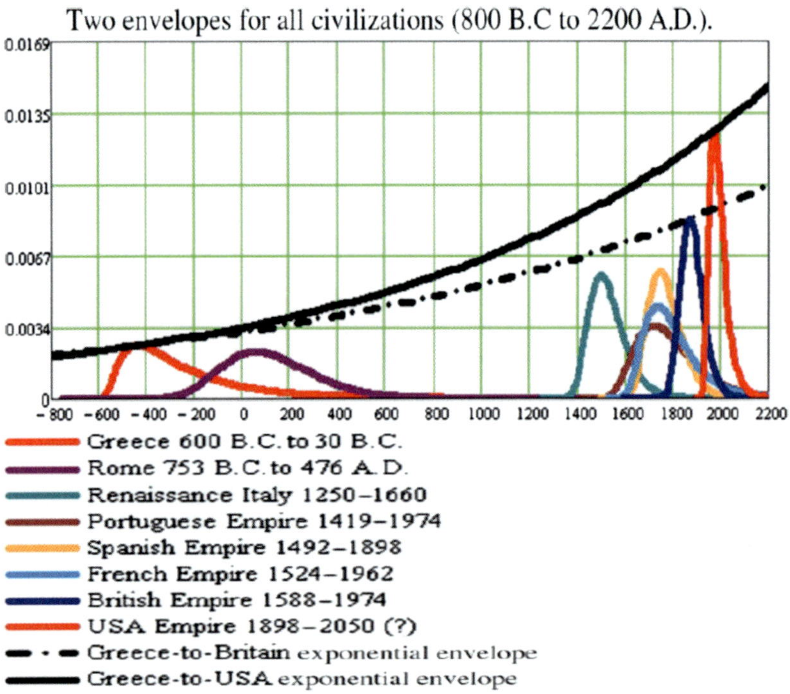

Finding the b-lognormals of eight among the most important civilizations of the Western world: Ancient Greece, Ancient Rome, Renaissance Italy, Portugal, Spain, France, Britain and the USA. For each such civilization three input dates are assigned on the basis of historic facts: (1) the birth time, b; (2) the senility time, s, i.e. the time when the decline began, and (3) the death time, d, when the civilization reached a formal end. From these three inputs and the two equations (57) the b-lognormal of each civilization may be computed. As a result, that civilization's peak is found, as shown in the last two columns. In general, this peak time turns out to be in agreement with the main historical facts.

	b = Birth time	s = Senility time	d = Death time	p = Peak time	p = Peak ordinate
Ancient Greece	600 BC Mediterranean Greek coastal expansion.	323 BC Alexander the Great's death. Hellenism starts.	30 BC Cleopatra's death: last Hellenistic queen.	434 BC Pericles' Age. Democracy peak. Arts and science peak.	2.488×10^{-3}
Ancient Rome	753 BC Rome founded. Italy seized by Romans by 270 BC.	235 AD Military Anarchy starts. Rome not capital any more.	476 AD Western Roman Empire ends. Dark Ages start.	59 AD Christianity preached in Rome by Saints Peter and Paul against slavery.	2.193×10^{-3}
Renaissance Italy	1250 Frederick II dies. Middle Ages end. Free Italian towns.	1564 Council of Trent. Tough Catholic and Spanish rule.	1660 1600 Bruno burned. 1642 Galileo dies. 1667 Cimento Academy Shut.	1497 Renaissance art and architecture. Science. Copernican revolution.	5.749×10^{-3}
Portugal	1419 Madeira island discovered.	1822 Brazil independent, colonies retained.	1999 Last colony Macau lost.	1716 Black slave trade to Brazil at its peak.	3.431×10^{-3}
Spain	1492 Columbus discovers America.	1805 Spanish fleet lost at Trafalgar.	1898 Last colonies lost to the USA.	1741 California to be settled by Spain, 1759–76.	5.938×10^{-3}
France	1524 Verrazano first in New York bay.	1815 Napoleon defeated at Waterloo.	1962 Algeria lost, as most colonies.	1732 French Canada and India conquest tried.	4.279×10^{-3}
Britain	1588 Spanish Armada Defeated.	1914 World War One won at a high cost.	1973 The UK joins European EEC.	1868 Victorian Age. Science: Faraday and Maxwell.	8.447×10^{-3}
USA	1898 Philippines, Cuba, Puerto Rico seized.	2001 9/11 terrorist attacks.	2050 ? Will the USA yield to China ?	1973 Moon landings, 1969–72.	0.013

5. Claudio Maccone then invoked Shannon's Information Theory[29]. The b-lognormals' entropy turns out to be the index of "development level" reached by each historic civilization. We thus get a numerical estimate of the entropy difference between any two civilizations, like the Aztec-Spaniard difference in 1519.

[29] In a landmark paper written at Bell Labs in 1948, Shannon defined in mathematical terms what information is and how it can be transmitted in the face of noise. What had been viewed as quite distinct modes of communication--the telegraph, telephone, radio and television--were unified in a single framework.

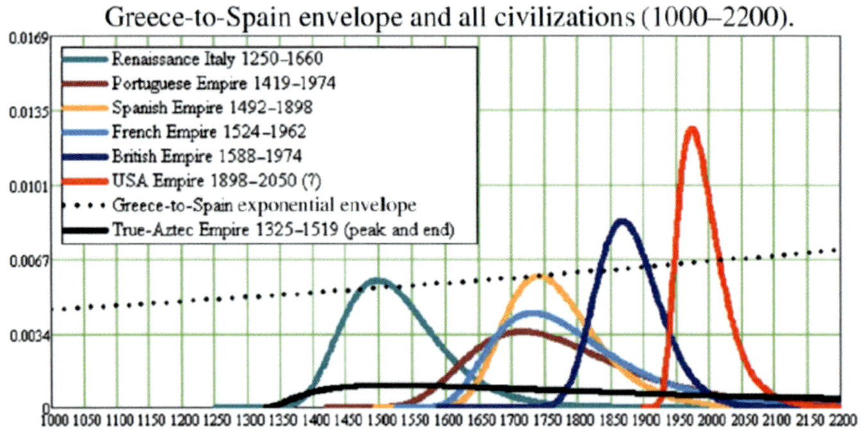

Other historic empires (for instance the Dutch, German, Russian, Chinese, and Japanese ones, not to mention the Aztec and Incas Empires, or the Ancient ones, like the Egyptian, Persian, Parthian, or the medieval Mongol Empire). Those historic-mathematical studies will be made at a later stage of development of this new research field that, in Maccone's view, is 'Mathematical History': the mathematical view of human history based on b-lognormal probability distributions.

To summarize this section's content, for each one of the eight civilizations listed above, he defines :

(a) Birth b, namely the year when that civilization was supposed to be 'born', even if only approximately in time.

(b) Senility s, namely the year of an historic event that marked the beginning of the decline of that civilization.

(c) Death d, namely the year when an historic event marked the 'official passing away' of that civilization from history.

Then, consider the two equations;

For each civilization, these two equations allow us to compute both μ and σ in terms of the three assigned numbers (b, s, d). As a consequence, the time of the given civilization peak is found immediately from the upper equation, that is

peak time = abscissa of the maximum = $p = b + e^{\mu - \sigma^2}$.

Also, we can then write down the equation of the corresponding b-lognormal immediately. The plot of this function of time gives a clear picture of the historic development of that civilization, though, to save space, we prefer not to reproduce here the above eight b-lognormals separately.

Inserting the peak time into the peak ordinate of the civilization is found, namely 'how civilized that civilization was at its peak,' and this is explicitly given by the lower equation, namely:

$$\text{peak\_ordinate} = P = \frac{e^{\frac{\sigma^2}{2}-\mu}}{\sqrt{2\pi}\sigma}.$$

The table that summarizes the three input data (b, s, d) drawn by the author from history textbooks, and then the two output data (p, P) of that Civilization's peak, namely its best legacy to other subsequent Civilizations.

(1) The first two civilizations in time (Greece and Rome) are separated from the six modern ones by a large, 1000 years gap. This is of course the Middle Ages, i.e. the Dark Ages, that hampered the development of Western Civilization for about 1000 years. Carl Sagan said, 'the millennium gap" in the middle of the diagram represents a poignant lost opportunity for the human species', Sagan (1980).

(2) While the first two civilizations, Greece and Rome, lasted more than 600 years each, all modern civilizations lasted much less: 500 years at most, but really less, or much less indeed.

(3) Since b-lognormals are pdfs, the area under each b-lognormal must be the same, i.e. just 1 (normalization condition). Thus, the shorter a civilization lives, the highest its peak must be! This is obvious from the diagrams: Greece and Rome lasted so long, and their peak was so much smaller than the British or the American peak!

(4) In other words, his theory accounts for the 'higher level of the more recent historic civilizations' in a natural fashion, with no need to introduce further free parameters.

This author was the first one who was able to give a quantitative description of both Darwinian evolution and human history, based upon his new discoveries about the mathematical properties of finite and infinite b-lognormals. He did this by introducing new mathematical techniques in evolutionary modeling.

3.2.2. The Inclusion of China into the model.

While Claudio Maccone in his paper takes over Carl Sagan's myth of the Medieval Gap and assumes that in this period the growth of human knowledge stagnated, the Chinese civilization developed four crucial inventions that would fuel the development and spreading of the Western capitalist civilization; the compass, the gunpowder, paper and the printing.

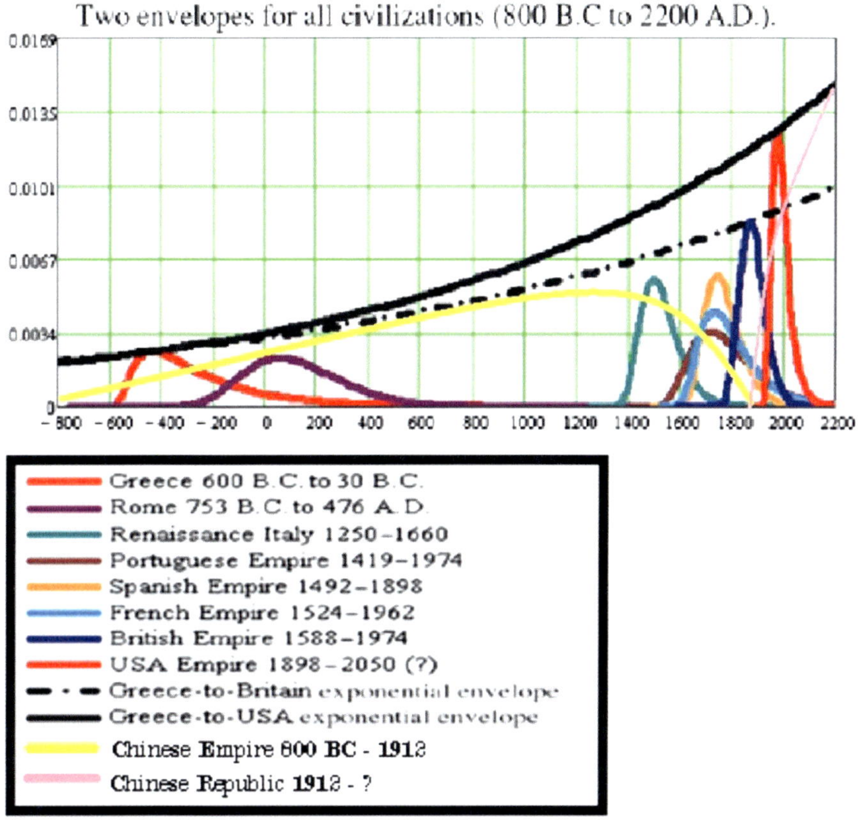

Karl Marx commented already in 1863 on the importance of gunpowder, the compass and printing, "Gunpowder, the compass, and the printing press were the three great inventions which ushered in bourgeois society. Gunpowder blew up the knightly class, the compass discovered the world market and found the colonies, and the printing press was the instrument of Protestantism and the regeneration of science in general; the most powerful lever for creating the intellectual prerequisites."[30]

John Needham, a British biochemist, historian and sinologist, published an extensive compilation of the East Asian contributions to pre-modern science, technology and medicine in which he also described the mechanisms of its transmission, thus demonstrating that modern science isn't exclusively rooted into the ancient Greek tradition. He illustrated this with a diagram that uses the terms transcurrent - and fusion points.

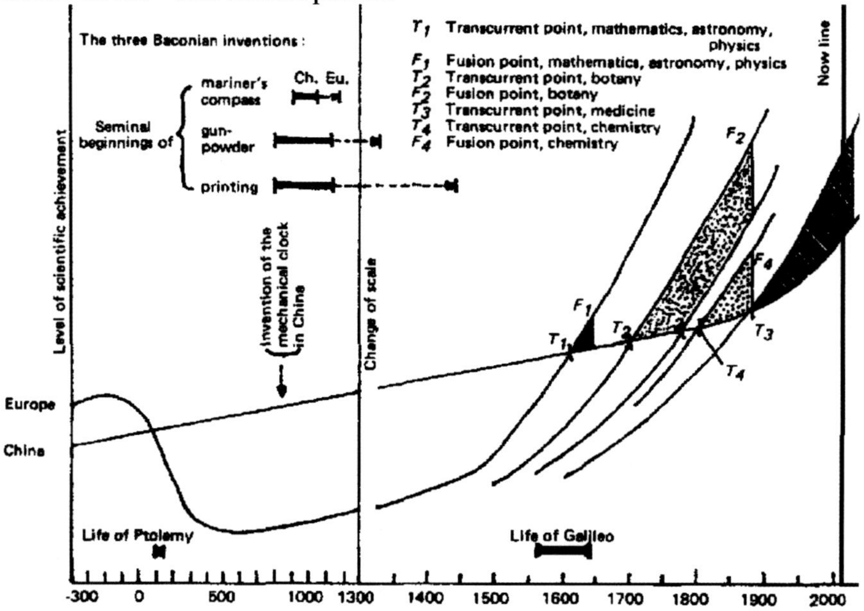

The "transcurrent point" designates that moment at which the European scientific and technical level surpassed that of a particular non-European civilization. In general, this moment would apparently occur simultaneously with or slightly after the successful investigation of any given field of study by means of modern methodology. The "fusion point" designates that moment at which the body of knowledge and technique belonging to a particular non-European traditional science is successfully incorporated into the modern scientific system.

What distinguished modern science from all the traditional sciences was its new methodology, that is, the practice of systematic experimentation to test mathematized hypotheses. In Needham's

---

[30] Marx, Karl. "Division of Labour and Mechanical Workshop. Tool and Machinery". Economic Manuscripts of 1861-63

opinion, traditional science typically framed its experimentation and analyses in a system of non-testable categories which were essentially regional or ethnic-bound. Modern science, on the other hand, used experimentation precisely for the purpose of axiomizing mathematically and of proving the validity of mankind's fundamental notions of reality; because of the introduction of mathematized propositions (which are universal, in the sense that they can be understood, tested, and developed by men and women of all nationalities), Needham calls modern science ecumenical.

One of the greatest needs of the world in our time is the growth and widespread dissemination of a true historical perspective, for without it whole peoples can make the gravest misjudgments about each other. Since science and its application dominate so much our present world, since men of every race and culture take so great a pride in man's understanding and control over her, it matters vitally to know how this modern science came into being. Was it purely a product of the genius of Europe, or did all civilizations bring their contributions to the common pool A right historical perspective here is one of the most urgent necessities of our time[31].

The Chinese empire used to be the result of a tension field between Confucianism and Taoist values. When the European colonizers arrived in Asia, the Chinese empire was already over its peak and descending into its senility phase, with an overly centralized government blocking or delaying the introduction of new ideas or technologies.

The contemporary Chinese republic is an amalgam of the tension field between communism, capitalism and traditional values. Into the diagram Needham drafted, he forecasts how the Chinese are fast closing in on a stagnating Western civilization. This is mainly done with utter disrespect for the notion of intellectual property and by taking the findings of Western science and technology to a higher level while waging an aggressive socio-economic expansion policy.

[31] Needham, 3. et al. (1954 -), Science and Civilization in China. CUP, in seven volumes. Vol. 1, 1954. Vol. II, 1956. Vol. III, 1959. Vol. IV, Part 1, 1962. Vol. IV, Part 2, 1965. Vol. IV, Part 3, 1971. Vol. V, Part 2, 1974. Vol. V, Part 3, 1976.

3.3. The Seven-Dimensional Nature of Culture

To understand culture is crucial for understanding human behavior. How culture comes to be and the division of labor between genes and culture in the determination of human behavior are central topics of this chapter that propagates an evolutionary approach of culture.

Unlike the predator/prey relationship where the predator requires the prey for success, with competition two populations would just as soon avoid each other all together.

This pressure towards avoidance is the source of much ecological diversity since it propels populations to explore new and therefore competition-free niches. An ecological niche, for some particular species, is simply that collection of resources the species relies on.

Interspecific niche overlap occurs when two or more species share one, some, or perhaps all of their resources. When those resources are scarce, interspecific competition will result. The width of a niche is simply a qualitative sense of the variety and number of resources a population makes use of.

The next paragraphs are intended to contribute to the clarification of the nature of these "guiding criteria" and find that high levels of competition occur more frequently among quasi-species of memes who exist within a narrow ecological niche. In the long term, the guiding criteria strongly bias which behaviors and skills enter the pool of cultural traits of the population, and as such they play a central role in determining the cultural evolution of a population.

In analogy with the 7-dimensional structure of the reality and of the human psyche, there exists a model that proposes a set of 7 different axis along which cultures differentiate one from another. Seven is a recurring number in the synthetically theory of the universe, humankind and religion.

Trompenaars and Hampden-Turner concluded that what distinguishes people from one culture compared with another is where these preferences fall in one of the following seven dimensions:

1. Universalism versus particularism; what is more important – rules or relationships?
2. Individualism versus communitarianism; do we function as individuals or as a group?
3. Specific versus diffuse; involvement, commitment and context. How separate do we keep or private and professional life?
4. Neutral versus emotional; do we display our emotions?
5. Achievement versus ascription; do we have to prove ourselves to be given status or is it given t us?
6. Sequential time versus synchronous time; do we do things one at the time or do we do several things at once?
7. Internal direction versus outer direction; strategy and planning. Do we control our environment or are we controlled by it?

You can use the model to understand people from different cultural backgrounds better, so that you can prevent misunderstandings and enjoy a better relationship with them.

The Trompenaars Hampden-Turner Seven Dimensions of Culture

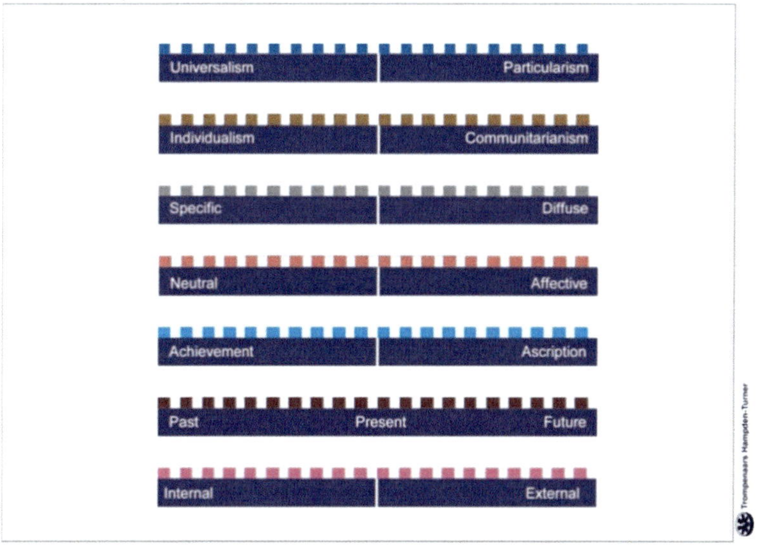

Alfred Marshall said we should use math when developing a new theory in order to reveal any hidden assumptions and make sure the argument is logically consistent, then throw out the math and present the argument in words.

New institutional economists are somewhat taken aback by the assertion that economists and some others in the social sciences relegate culture to the role of a minor player in the determination of human behavior.

Though new institutional economists assume that preferences have a genetic component, everything else that influences our behavior is assumed to be a product of the interplay between scarcity and culture, but economists often ignore the fact that cultural evolution is a path dependent process and that individuals from different cultures will respond to the same economic incentives in different ways.

3.4. The Synthetic Approach of Literature.

3.4.1. A Multidisciplinary Angle.

There really are no such things as Art or Science. There are only artists and scientists. Take as an example all those discussions about Leonardo; his art and science. For Leonardo, art was a skill, a know-how applied both to his scientific experiments and to painting. I go back deliberately to the old meaning of the term "art," when art was identified with skill or mastery - the art of war, the art of love, or whatever else. Art is something with a skill. There's no disembodied skill as such; skill is always applied to a particular task.

Art has to do with the embodiment of our value systems: we value elegance and tenderness; love and other such emotions are ingredients in our value system. I do think that the intellectual level of discourse of, say, a number of great scientists and Nobel Prize winners, is much higher than the level of discourse in most art criticism. Much of the activity of art historians has been wedded to whatever is fashionable in the art market at a given time. There is a constant temptation for artists to be sensationalists - not to make great works of art or even minor works of art but merely talking points - and to achieve notoriety for a time.

Artistry can be defined as having mastered a skill sufficiently enough so that you don't have to think about it; you live it. Artistry is the bridge between concept and craft. Once you have mastered a skill you can transcend technicalities and focus on creating, inventing and innovating. Artists and scientists constantly work their crafts by developing their skills. In order to take on more challenge and stay in the flow, you constantly need to learn new skills. Mastery is what separates the virtuoso from the technician; in science, art and living.

A contemporary illustration is the work of Alexander Rodin. I stumbled upon his paintings in 2011, during an exposition called East Meets West in Kunsthaus Tacheles, Berlin, Germany. Rodin currently lives and works in Berlin, because (as he describes it himself in an interview with website n-europe.eu) 'Berlin is a place interested in art. [...] I am an artist and exhibitions are my life. And when I suggest a concept of an exhibition or an event and see that it fascinates Germans – why not to organize it there? Why not to hold an exhibition in Berlin when it acknowledges my creativity?

Rodin's most inspiring works are rarely for sale, but the artist gladly sells signed posters with reproductions of his works or puts them up for expositions. That's how he distillates a living from his art.

Alexander Rodin's paintings are an event quite unexpected for Belarusian art with its ideological taboos and conventional norms. The art composition in his paintings is based upon the multifaceted reflections of infinite spaces. The latter undergo constant changes of urban, geological, technocratic and biological nature. And in order to take them to a single image-bearing system Alexander chooses a high panoramic point of view and uses practical monochrome colors of various tonalities. Alexander paints the universe with microscopic precision. The longer one peers at those silver-black abysses the deeper they become...

He describes his works as follow, "I work in a manner that's unfortunately moved to the background now; it has been substituted for coloring and graphics. I combine avant-garde forms and introduce constructivism and cosmogony and through the plasticity of the rhythms and color structures I try to juxtapose the micro and macro cosmos of mental, physical and astral being".

Art and science are the results of actions by people who're living while paying attention

It connects with my own research about literary criticism that I've called "Synthetic Literary Criticism". This method is intended to indicate a concrete, positive description of moving equilibriums, oscillations, and secular change, by a method which presents all of the interrelated literary quantities in a synthesis of simultaneous, real equations.

From Euclid of Alexandria till the late 19th century was it a quite common phenomenon for philosophers to be also practicing mathematicians. The split-brain only occurred when the modernist thinking founded itself, sometime in December 1910, upon the inherent instability of the new relativist worldviews.

Recently there is a renewed interest in some artistic circles to restore the unity between the different disciplines of knowledge; to create a new holistic worldview.

Some celebrated writers did resort to hidden or explicit mathematical structures to construct novels that are now widely regarded as literary masterpieces.

The first example includes Georges Perec who, in his novel "Life– a User's Manual", takes us on a tour, chapter by chapter of the apartments (stages of the lives) of the tenants of an entire Parisian building in a sequence purely determined by graph theory.

Another instance of such approach is that of the influential novel “Hopscotch” (Rayuela), where Julio Cortazar invites the reader to navigate through his complex puzzle of 155 chapters by choosing either a linear or a non-linear mode of reading, each route yielding of course vastly different perspectives.

Jan Potocki’s superb "Manuscript Found in Saragossa", also known as "The Saragossa Manuscript" where the chapters are devoted to the geometer's adventures contain some real mathematical problems, usually associated with Pascal.

Jorge Luis Borges, a dedicated amateur-mathematician, was much intrigued by Zeno’s paradox and the concept of infinity which

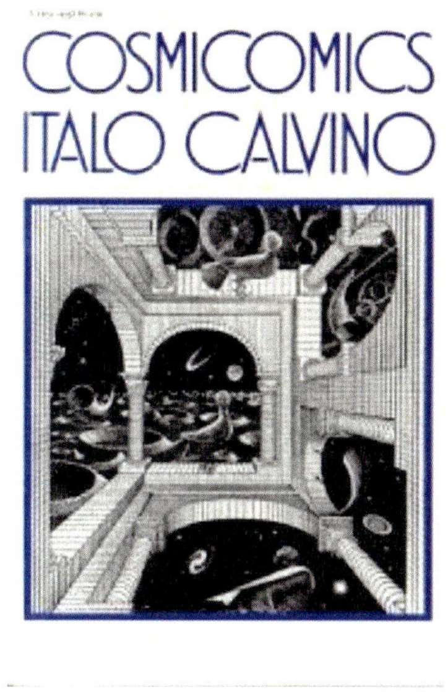

inspired a number of his stories, like the well-known “Library of Babel” or the “Book of Sand”.

The Italian writer Italo Calvino, in particular the "Cosmicomics" and the "Ti zero" uses mathematical theories or scientific facts as the starting point to create a complete world of characters that are "entities" which interact following the law of the universe.

A noteworthy newcomer is code poetry; literature that mixes notions of classical poetry and computer code. Unlike digital poetry, which prominently uses physical computers, code poems may or may not run through executable binaries. "Black Perl" is a code poem written using the Perl programming language. It was posted anonymously to Usenet on April 1, 1990 and is popular among Perl programmers as a piece of Perl poetry. Written in Perl 3, the poem can be executed as a program.

```
BEFOREHAND: close door, each window & exit; wait until
time.
    open spellbook, study, read (scan, select, tell
us);
write it, print the hex while each watches,
    reverse its length, write again;
    kill spiders, pop them, chop, split, kill them.
        unlink arms, shift, wait & listen (listening,
wait),
sort the flock (then, warn the "goats" & kill the
"sheep");
    kill them, dump qualms, shift moralities,
    values aside, each one;
        die sheep! die to reverse the system
        you accept (reject, respect);
next step,
    kill the next sacrifice, each sacrifice,
    wait, redo ritual until "all the spirits are
pleased";
```

```
    do it ("as they say").
do
it(*everyone***must***participate***in***forbidden**s*e
*x*).
return last victim; package body;
    exit crypt (time, times & "half a time") & close
it,
    select (quickly) & warn your next victim;
AFTERWORDS: tell nobody.
    wait, wait until time;
    wait until next year, next decade;
        sleep, sleep, die yourself,
        die at last
```

Code can speak literature, logic, mathematics. It contains different layers of abstraction and it links them to the physical world of processors and memory chips. All these resources can contribute in expanding the boundaries of contemporary poetry by using code as a new language. Code to speak about life or death, love or hate.

It has its own rules (syntax) and meaning (semantics). Like literature writers or poets, coders also have their own style that include - strategies for optimizing the code being read by a computer and facilitating its understanding through visual organization and comments for other coders.

The new information technology added more extra dimensions to the literary field. Social media networks, especially Twitter and Tumblr,

have proven to be hospitable environments for functional and artistic bots because they operate as publishing and development platforms that encourage automation and produce massive data streams.

Social bot accounts (Sybils) have become more sophisticated and deceptive in their efforts to replicate the behaviors of normal accounts.

The term “Sybil” comes from the subject of the book Sybil (a woman diagnosed with dissociative identity disorder). Bots can be designed for good intentions. They can be used to protect anonymity of members as mentioned in related work or automate and perform tasks much faster than humans, like automatically pushing news, weather updates or adding a template in Wikipedia to all pages in a specific category, or sending a thank-you message to your new followers out of courtesy. They can also be extremely sophisticated such as (i) generating pseudo posts which look like human generated to interact with humans on a social network, (ii) reposting post, photographs or status of the others, and (iii) adding comments or likes to posts, (iv) building connections with other accounts.

But as many technological inventions, they also have darker applications. Some of the malicious functionalities of social bots are: the power of dissemination of misinformation, they are convenient way of propaganda or can be leveraged for getting fake ratings and reviews.

There are influence bots that serve this purpose. Also, it is possible to find many web pages that serve fake followers and likes even for free by simply searching on any search engine. Social networks are powerful tools that connect the millions of people over the world. Therefore, they are attractive for social bots as well. The possible harm caused by social bots such as identity theft, astroturfing, content polluter, follower fraud, misinformation dissemination etc… may not be underestimated.

Figure 6: An industrial robot writes out the Bible. Picture by Amy Cicconi

Is there greater cause to worry further up the literary food chain since the rumor circulates that GPT-2 bots can also write books?

No doubt Amazon would jump at the prospect of being able to sell completely computer-generated books since they would not need to pay royalties to an algorithm.

Up to now, these are just giant automated plagiarism machines that mash together bits of stories written by human beings rendered invisible by AI rhetoric. The major GPT-2 glitch consists of the fact that it can be sometimes prone to what its developers call “world-modeling failures”, “eg the model sometimes writes about fires happening underwater”.

A more realistic hope for a text-only program such as GPT2, meanwhile, is simply as a kind of automated amanuensis to generate the elusive raw material that human writers can then edit and polish. But until robots have rich inner lives and understand the world around them, they will not be able to tell their own stories. And even then.
Maybe there is no better way to conclude this introduction than by quoting Wittgenstein: “If a lion could speak, we would not understand him”.

3.4.2. Literary Criticism and Mathematics

Let me start by acknowledging that there exists inside the literary community a strong resistance against the mathematization of their field of interest. Take the rejected master's thesis of Kurt Vonnegut at the University of Chicago between 1945 and 1947 about the six basic plots in literature wherein he stated that “stories have shapes which can be drawn on graph paper and that the shape of a given society’s stories is at least as interesting as the shape of its pots or spearheads”.

As far as many bookworms are concerned, advanced equations and graphs are the last things which would hold their interest, but there is no escape from the math.

Physicists from the Institute of Nuclear Physics of the Polish Academy of Sciences in Cracow, Poland, performed a detailed statistical analysis of more than one hundred famous works of world literature, written in several languages and representing various literary genres.

To convert the texts to numerical sequences, sentence length was measured by the number of words … and discovered that fractals are everywhere; Joyce, Proust, Cortázar, Woolf, Dos Passos, Bolaño—fractals. Some of the world’s greatest writers appear to be, in some respects, constructing fractals.

However, more than a dozen works revealed a very clear multifractal structure, and almost all of these proved to be representative of one genre, that of stream of consciousness. The only exception was the Bible, specifically the Old Testament, which has so far never been associated with this literary genre.

"It is not entirely clear whether stream of consciousness writing actually reveals the deeper qualities of our consciousness, or rather the imagination of the writers. It is hardly surprising that ascribing a work to a particular genre is, for whatever reason, sometimes subjective. We see, moreover, the possibility of an interesting application of our methodology: it may someday help in a more objective assignment of books to one genre or another," notes Prof. Drozdz .

The most common mathematical approach of literature is of a statistical nature and is used to quantify the different aspects of a work of literature. That way some linear models could be distilled to reproduce certain literary phenomena. Binongo & Smith wrote on this subject “The availability of computing devices and the proliferation of electronic texts (the so-called 'e-texts') in centers for literary and linguistic computing in major universities have encouraged non-traditional applications of statistics. With the drudgery of computation and text encoding diminished, research in the field of computational stylistics is accelerating. In it is shown how projections onto the Cartesian plane of 25-dimensional vectors related to the frequency of occurrence of 25 prepositions can distinguish between Oscar Wilde's plays and essays. Such an application illustrates that it is possible to find unusual and intriguing examples of how statistics can impinge on unexpected territory”.[1]

The second approach to literature is the structuralism approach that was defined by Simon as “the belief that phenomena of human life are not intelligible except through their interrelations. These relations constitute a structure, and behind local variations in the surface phenomena there are constant laws of abstract culture"[2]. Structuralisms believe that everything that is written has to be governed by specific rules, or a "grammar of literature", that one learns in educational institutions and that has to be unmasked[3].

Both approaches fail to provide a clear conception of the interdependence of the parts of the literary system where all parts are connected and react to each other. It seems, therefore, as if, for a complete and rigorous solution of the problems relative to some parts of the literary system, it was indispensable to take the entire system into consideration.

In this essay I would like to defend the concept that no general method for the solution of questions can be established which does not explicitly recognize, not only the special numerical bases of the science, but also those universal laws of thought which are the basis of all reasoning, and which, whatever they may be as to their essence, are at least mathematical as to their form.

3.4.3. A Brief Contextual Synopsis of the US Literary Canon.

In the early days of independence, American novels served a useful purpose. They used realistic details to describe the reality of American life. But when some of the good American literature started to arise above the time and place where they were written; these works became universal.

The oldest examples are the sketches and observations of Michel Guillaume Jean de Crèvecoeur (1735 - 1813), bundled in “Lettres d'un cultivateur Américain”, and published in Paris, 1787. He did not describe America as a Utopia, nor did he expect it to become one, but he saw more hope and health in a society where: "individuals of all nations are melted in a new race of man" than in the older, closed societies of Europe.

Modern Chivalry by Hugh Henry Brackenridge was the first important American novel. He wanted to achieve a reform in morals and manners of the people. The book is a series of adventures in which the author laughs at America’s backwards culture.

Charles Brockden Brown's interest in the psychology of horror greatly influenced such writers as Hawthorne and Poe many years later. In his mayor novel “Wieland”, things may not be as they appear, and genuine truth must be actively searched for. With this philosophy, it is not surprising that he spent his last years with political pamphlets against the optimistic philosophy of Jefferson.

1810 - 1840 is known as the Knickerbockers’ Era of American literature. The name comes from "A History of New York" by Dietrich Knickerbocker, a pseudonym of Washington Irving. It was a humorous rather than a serious history of the city. He invented many of the events and legends he wrote about in the book. “The Sketch Book" (1819) contains two of the best loved stories from American literature, The Legend of Sleepy Hollow & Rip Van Winkle. It was based on old German folk tales filled with the local color of New York's Hudson River Valley. Irving regarded the story as "a frame on which I sketched my materials". Neither Irving nor any of the other Knickerbockers really tried to speak for the whole country. For them the world tended to stop at the borders of the New York State.

The victory of time and civilization is beautifully described by J.F. Cooper. Natty Bumppo appears in all of his novels and is one of the best-known characters in American Literature. Nathaniel “Natty” Bumppo was a child of white parents who grew up among the Native Americans. He criticized the wastefulness embodied in the settlers and demonstrated a way of life that was a synopsis of man and nature in the West. Also, in Bryant's Thanatopsis (North American Review, September 1817), the life of man is part of the nature as whole, and death is the absolute end of the individual.

In the 1830 - 1840's the frontier of American society was quickly moving to the West. At this time, Boston and its neighboring towns and villages were filled with intellectual activities. In the center of these activities were the Transcendentalists. The Transcendentalists tried to find the truth through feeling and intuition, rather than through logic. In their vision “Wisdom does not inspect, it beholds" (Thoreau). Thoreau's most famous book was Walden (1854). Apparently, it speaks only of the particular side of living alone in the woods, but in fact it is a completely Transcendentalist work. He was convinced that while civilization has been improving our homes, it has not equally improved those who live in them. He observed that the mass of humankind lead lives of quiet desperation and wrote on that subject; “As if you could kill time without injuring eternity".

Emerson stated in a publication (Nature) that man should not see nature merely as something to be used; that man's relationship with nature transcends the idea of usefulness. First Emerson would "deposit" ideas in his journal (which he called his bank account) and then he developed his lectures from the notes in his journal. Self-Reliance (1841); to believe in your own thought, to believe what is true for you in your heart is true for all man, - that is genius. He believed that to be great is to be misunderstood and considered a foolish consistency as the hobgoblin of little minds. In "The Over-Soul” (1841) he claimed that “We live in succession, in division, in parts, in particles. Meantime within man is the soul of the whole; the wise silence; the universal beauty, to which every part and particle is equally related, the eternal ONE. And this deep power in which we exist and whose beatitude is all accessible to us, is not only self-sufficing and perfect in every hour, but the act of seeing and the thing seen, the seer and the spectacle, the subject and the object, are one. We see the world piece by piece, as the sun, the moon, the animal, the tree; but the whole, of which these are shining parts, is the soul" (cit.).

European observers, who take a close look at the characteristics that they qualify as typical for white Americans, discover soon that they were originally attributes of the American Indian. The liberation from a social hierarchy and the idea that "all men are born equal" is also an American Indigenous invention that crossed the Atlantic Ocean and made old European feudal institutions crack at their foundations.

Not that the white American will give credit to the indigenous tribes for those values since they are still perceived as a hostile and primitive culture. This attitude can be held responsible for much of the dissatisfaction and restlessness found in the U.S. To trace down the roots and evolution of the American mentality, there is no better source than to follow the historical development of the American Literature. All along the attentive reader can feel the mostly invisible presence of the Native American and the subtle incorporation of his values in the American mainstream of thinking.

Poe's method was to put his characters in unusual situations. Next, he would describe their feelings of terror or guilt. Poe was also one of the creators of the modern detective -story. Instead of examining characters and feelings, these stories examine mysteries or problems in an attempt to liberate the reader from cultural conformity.

Longfellow was an excellent linguist and gained a lot of popularity with poetry containing pseudo-profundities as "Life is real and life is earnest, and the grave is not its goal". Longfellow borrowed legends of colonial times but his main contribution to the American culture was to translate European poetry and make it accessible for all Americans. Where Longfellow was more a facilitator, Brett Harte (1836 - 1902) wrote original stories about the Far West and many writers followed his lead. In all Harte's work we see all the main characters of the West American folk culture return; the pretty New England schoolteacher, the sheriff, the bad man, the gambler and the bar girl.

Mark Twain’s writing was strongly influenced by his work as a pilot on the Mississippi. He became nationally famous with "The celebrated jumping Frog", based on stories he heard as a journalist in a mining camp. Twain's work is filled with stories about ordinary people tricking experts. In "The adventures of Tom Sawyer", his two heroes are "bad boys" because they fight against the stupidity of the adult world. Some critics complain that he wrote only well when he was writing about young people. Throughout all of Twain's writing we see the conflict between the ideals of Americans and their desire for money.

By 1875, American writers were moving toward realism in literature. William Dean Howells (1837 - 1920) stated that romanticism created false views about life. Like most Americans in the 1880's, he realized that business and businessmen were at the center of society, and he felt that novels should depict them. Later he began attacking the evils of American capitalism.

In the 1890's, many realists became naturalists, a term created by Emile Zola. For them realism was an ideology and the novel had the power to become a political weapon. Crane had the view that life and death are determined by fate. He wrote about a man who said to the Universe "Sir, I exist! “
“However," replied the Universe "The fact has not created in me a sense of obligation".

Thanks to modern psychology and writers like Henry James, we are now more interested in the working of the human mind. We know that events inside one's head can be as dramatic as events in the outside world.

Adams is best remembered for his "St Michel and Chartres". On the surface it is a guidebook to two famous French religious sites. However, it is a deep study of medieval culture. The old Europe had a calm unity; the new culture of America, however, had neither calmness nor unity. One of Adam's finest quotes was; "Chaos is the Law of nature, order, the dream of man".

Meanwhile on the Old Continent at the turn of the 19th century, hypocrisy was the cement that kept the society together, cumulating in a nostalgic enjoying of decay that regretted only the loss of sexual opportunities. Art Nouveau wanted to bring the heaven to earth, but created only artificial paradises separated from the cities, hidden under their streets or on elevations above them. They were closed reservations of esoteric symbolism and pseudo-ritualism; the idioms of a retreat in the inner self.

The painting "Trismegisturian Harlequins" (Picasso 1901 -1905) portrayed the neglected offspring of Hermes. Because evolution in art or science is often the result of disrespect and revolt, they were punished with social exclusion for their diverging talents. Simmel and Durkheim considered that the collective always cannibalizes the individuality till it becomes socially acceptable. Freud declared in 1906 that the masochistic satisfaction that people experience when a theatre character suffers on stage, originates from the primitive offer rites who suppressed the impulse to revolt. He also stated that love is no remedy against war because sex and death aren't enemies but clandestine collaborators.

The Newtonian space and time collapsed in 1905 when Albert Einstein, at the age of 26, published four groundbreaking papers; On the Photoelectric Effect; Brownian Motion; The Special Relativity, and The Equivalence of Mass and Energy. The new concept is that all objects are moving vibrations in the space-time continuum. The quantum physician Heisenberg introduced the notion that the act of observing changes the observed object. The world was thought in pieces, just as the ego, which is perceived as just a particle in an unlimited and unprotected field.

On December 11th, 1910, the human character changed, and the modernist consciousness founded itself upon the inherent instability of the new relative worldviews. Suddenly humankind was lost in a relativistic universe where there are no more rules that can be transgressed, and nobody can be accused of abnormality. The Chinese pagoda became popular in European culture as the cathedral of a religion that accepts a void because its five floors symbolize the five mystical nothings of life (earth, water, air, fire and ether) and the reality of a barrel became its emptiness inside (Lao-tse).

Existentialism placed the individual on a rope that spans an abyss and turned his daily occupations in acts of courage and perseverance. Language became, just as money, an exchange object whose circulation serves only as a vector that maintains a sense of community. Wittgenstein perceived the language as the limit of our knowledge and declared; „Wovon man nicht reden kann, darüber muss man schweigen“. But the modernist thinking crosses the language barriers and stipulates that the unspeakable can be expressed in paintings, music or mathematical equations.

Another modernist credo is that the artist or scientist cannot create anything new because the world is already made. All they can do is to reorganize it and being innovative is to force the reality in new molds of perception. The modernist art diverges in two directions; the impressionists who wait for the evaporation that will drain away the observable world in favor of the pure and unlimited space and the cubists who replaced religion by a happily chaotic pantheism. When Apollinaire in 1913 analyzed a cubist painting, he warned for the hybrid and deviating creatures that modernism can produce. He predicted a world were individualism will be erased and humankind mentally and physically adapted to serve the machines.

The most intriguing paradox of the 20th century modernists was their tendency to primitivism; the desire to abandon the laboriously earned advantages of culture and science in order to be attracted by a reckless regression. This fanatically cultivated barbarism elevated actors, musicians and athletes to the status of temporary deities to be worshiped by the masses. The Rite of Spring by Stravinsky became the musical expression of modernist’s primitive nature. It opposed the social contract that binds the individual to an institutionalized and repressive conscience.

The human condition still has to come to terms with the idea that primitivism and industrialism are allies in the modernist vision. The dilemma the of the 20th century is synthesized by their affinity; the primitive culture and art of the American Indian on one side, and the worldwide atrocities committed on the battlefields on the other side, show the world before and after the mechanization. So is it not astonishingly that most of the contemporary American isolationism originated from regions with the closest ties to the indigenous populations.

American newspapers were becoming very powerful by this period. They were very patriotic and pleased their readers with stories of courage and red blood. American society was united in a kind of conspiracy against the growth and freedom of the mind.

For too long, American life had been divided between the businessman (who only thinks of making money) and the intellectual (who has only unpractical theories and ideas).

The new generation of American writers did construct a middle ground where they meet. Hiding the truth about human sexuality - and punishing those who tried to talk about it - was part of America's puritanical morality. Only intelligent readers were able to accept even ugly truths about human nature.

This can be illustrated by a quote from Edith Warter who wrote about a ladies' culture circle in Xingu (1916) "Her mind was like a hotel where ideas came and went like transient guests, without leaving their addresses behind". Edith Warter was brought up to see herself as a decorative object for wealthy man. The upper class claimed to be highly moral, but often, their actions - towards women as well as in business - were not moral at all. In all her works, the natural instincts of people are crushed but an untruthful society

Theodore Dreiser's characters did not attack the nation's puritanical moral code; they simply ignored it. While looking at individuals with warmth, he also sees the disorder and cruelty of life in general.

Sherwood Anderson brought the techniques of modernism in American literature. These techniques include; a simpler writing style, more emphasis on the form of the story than on its content and a special use of time.

Sinclair Lewis character, Babbitt, is part of the American language. It means a joiner, a conformist. Lewis severely condemns the values

of middle-class America, but he doesn't suggest any other values that can take their places.
Ernest Hemmingway was trying to create his own answer to that problem. Many of the young people in the post-war I period had "lost" their American ideals.
Fitzgerald hero Gatsby symbolizes the belief that money can buy love and happiness. His failure makes him a rather tragic figure. In this context, Hemingway spoke about a lost generation. Without hope or ambition they tried to enjoy each day as it comes. All they want to know is how to live in the emptiness of the world. The typical Hemingway hero must always fight against the nada of the world. He developed a theme of heroism, stoicism and ceremony.
In the early thirties, the first reaction to the depression was a literature of social protest. The failure of the American dream became the main theme in Jewish-American literature. The novel “Call it Sleep" mixes Marxism and Freudian theory, Jewish mythology and a stream of consciousness writing style. Farrell writes more about spiritual poverty then about economic poverty in "Young Lonigan"(1934). It handles about the emotional religion, the new child every year, the money-worries and the heavy drinking of Irish-Catholic families.
The work of John Steinbeck represents a similar attempt to get it all on paper. In "The Grapes of Wrath" (1939) he told the story of a great national tragedy through the experiences of a single family of individuals. A couple of years later Miller (1891 - 1980) called America "an air-conditioned nightmare". In "The Chronological Eye" (1939) he stated that all his life he had felt a great skin ship with the madman and the criminal. He developed his own vision of how man should live. Laughter, freedom and joy should be the goals of life.
By 1945, America was a world power with huge international responsibilities. After the war, America entered in an" Age of Anxiety". First there was the fear of The Bomb and The Fear of Communism became a national sickness. Many writers in this period tried to find answers to old questions like "Who am I?” They tried to find it in their own racial backgrounds, while others explored the new ideas of philosophy and psychology and the young beat-writers used oriental religion for the same purpose.

In the 1940's 1950 the Jewish-American literature looked at the spiritual and psychological problems in a different way. They brought a new interest in the old moral problems and created the humor of self-criticism.
Saul Bellow once stated, “According to the philosophy of existentialism, man is completely alone in a meaningless world without God or absolute moral laws”. The Bellow-hero lives actively in his own head. He searches for answers in his mind rather than in the outside world while Singer wrote about eastern-European Jews, their superstitions and folktales and brought this lost world back to live.
Norman Mailer did more than trying to describe the existential pain of modern world. "I will settle for nothing less than making a revolution in the consciousness of our world". He reported on the psychological history of America while that story still happened. Some other writers in the sixties and seventies looked deep into the nature of American values in order to understand what is happening in their souls.
The characters in Updike's later novels seemed to have only their bodies. These bodies became more important than their souls. In his books, Updike becomes the novelist of the modern religion; sex.
Important new forms for American fiction were;
1. Factualized novel; the author used the facts of history to create new and unusual forms of fiction.
2. Post-realism; we can no longer be sure that there is a "real world" outside our heads.
In Vonnegut's novels, life was described as a terrifying joke. Real time was broken up into little bits and mixed together. Then he turned away from experimentalism, towards a style where his humor was still black, but softer and less painful. This attitude is equally reflected by Barth who wrote, “What the hell, reality is a nice place to visit, but you wouldn't want to live there, and literature never did, very long".
Vladimir Nabokov agreed and believed that art is a kind of reality where "the invention of art contains far more truth than life's reality". In his opinion, "Fiction is the most urgent game, a contest of minds with the reader". Nabokov was an artist who tried to "defeat time and destroy reality".

In Brautigan's book “In Watermelon Sugar" a character asks the narrator; "What's your book about?”. He answered, "Just what I'm writing down, one word after another".
This contrasts with the stories of Thomas Pynchon, whose plots and the things he wrote about were mostly real. He was unusual because he seemed to know everything. His novel "Gravity's Rainbow" has been studied by Scientific American because of the interesting ideas in it. Pynchon's novels tried to create the "emotion of mystery". His main characters became detectives, spending their lives to understand strange mysteries. One of the leading themes of his novels seems to be "What comes now?”

An often-neglected facet of US literature is science fiction, a genre that contemplates possible futures. Because science fiction spans the spectrum from the plausible to the fanciful, its relationship with science has been both nurturing and contentious.

HG Wells who, by most critics, is considered the US pioneer of the genre, used his time machine to take the reader to the far future to witness the calamitous destiny of humanity.

The renowned novelist and poet Ursula K. Le Guin said once, “The future is a safe, sterile laboratory for trying out ideas where anything at all can be said to happen without fear of contradiction from a native… a means of thinking about reality, a method.” Her award-winning 1969 novel, The Left Hand of Darkness—set on a distant world populated by genetically modified hermaphrodites—is a thought experiment about how society would be different if it were genderless.

William Gibson published in the 1980s the cyber punk novels wherein he depicted visions of a hyper-connected global society where black-hat hackers, cyber war and violent reality shows are part of daily life. He also coined the term cyberspace.

In recent decades there is a tilt toward dystopian futures, partly because of a belief that most of society has not yet reaped the benefits of technological progress. Bertrand Russell’s words from 1924 are prophetic when he wrote: “‘I am compelled to fear that science will be used to promote the power of dominant groups, rather than to make men happy.'

This fear is shared by Kim Stanley Robinson—the best-selling author of the Mars trilogy, 2312 (2012) and Suzanne Collins' who authored The Hunger Games (2008), in which a wealthy governing class uses ruthless gladiatorial games to sow fear and helplessness among the potentially rebellious, impoverished citizens.

This tendency can be seen as a countermeasure to the future shock that will become more intense with every passing year. The mathematical model of US literature demonstrates that the subject line of the bestselling US fiction is flat lining since 1965, which might be another indicator that the US civilization is over its peak and might have entered its senility phase.

3.4.4. The Universe and the US Literary System.

Since literature has the whole universe as a subject, one could assume that this system is also governed by the same mechanisms that it describes.

UNIVERSE

Alan Guth believes that the universe bubbled up out of a pre-universal singularity[32]. During a short moment, all the forces and building stones of matter were one. When the Higgs-field symmetries started to break up, followed a hot expansion [33]. For the function of our actual model of the Big Bang, the particles of our universe must have been organized with such a perfection that cannot be explained by coincidence. This improbable circumstance is known as the homogametic principle.

When the Higgs-field instead of exploding, coagulated, can it be possible that this created parallel universes, divided by energy fields that are dividing the universe into different domains [34].

The Universe is governed by four fundamental forces; Gravity, Electromagnetism, Weak Core Energy and Strong Core Energy with the space-time[35] as background.

[32] *Guth, Alan H (1997), The Inflationary Universe, Reading, Massachusetts: Perseus Books, ISBN 0-201-14942-7*

[33] *Higgs, Peter 1964. Broken symmetries and the masses of gauge bosons. Physical Review Letters 13 (16): 508–509*

[34] *Josh Clark (1998-2014). "Do parallel universes really exist?". HowStuffWorks website.*

[35] *Hermann Minkowski, "Raum und Zeit", 80. Versammlung Deutscher Naturforscher (Köln, 1908).*
Physikalische Zeitschrift 10 104-111 (1909) and Jahresbericht der Deutschen Mathematiker-Vereinigung 18 75-88 (1909). For an English translation, see Lorentz et al. (1952)

LITERATURE

Frye launched the idea that the mythological framework of American Literature and Cosmology is provided by the Bible. He developed this theory in his work "The Great Code"[36] in which he established how the Bible's narrative is related to all conventions and genres of American Literature. In this context, we assume that the pre-literary singularity of the American Literary Model is the Bible. The first Western settlers arrived around 1561 on the American continent and the Bible or - related literature stood at the center of their existence till 1716 (see table I of previous chapter). When these first symmetries started to break up, followed a hot expansion of the American Literary Universe. We postulate that time in American Literature started with this Big Bang and that previous times are undefined.

In this essay we consider the literature of other times, continents and cultures as parallel literary models.

The most influential parameters of the literary system are subject, language, protagonists and style with a socio-cultural time frame as background. The terms 'centripetal' and 'centrifugal' are used to categorize these parameters. A parameter is essentially centripetal when it moves outwardly, away from

[36] *Northrop Frye. The Great Code: The Bible and Literature. New York: Harcourt Brace Jovanovich, 1983.*

UNIVERSE

Our understanding of the cosmological horizon sets also the scale of the universe we can observe. In this sense, future events could already influence the present, depending upon the scale of the applied cosmically horizon[37].

LITERATURE

the text to the outer world and language are two centrifugal parameters while protagonists and style are centripetal

All works of literature have an origination, an escalation of conflict or paradigm, and a resolution. It works itself through this cycle by going through five phases; orientation, crisis, escalation, discovery and change. Those phases can come in following sequences: linear, nonlinear, interactive narration and interactive narrative.

---

[37] *Margalef-Bentabol, Berta; Margalef-Bentabol, Juan; Cepa, Jordi (8 February 2013). "Evolution of the cosmological horizons in a universe with countably infinitely many state equations". Journal of Cosmology and Astroparticle Physics. 015. 2013 (02).*

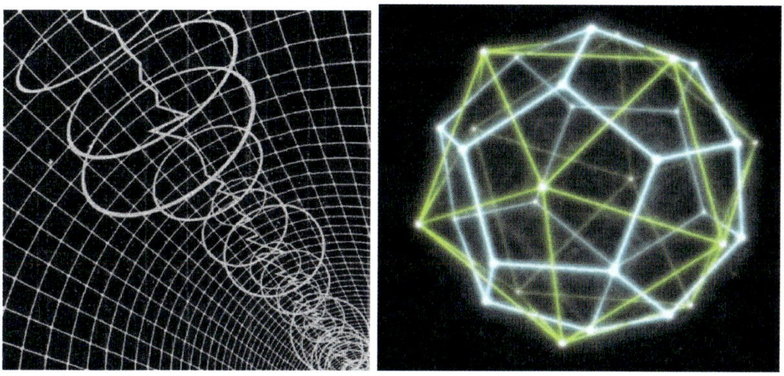

To the left a visualization of the evolution of a centrifugal function through the literary system with the centripetal functions circling around it. To the right a detail of multiple centripetal functions circling around a centrifugal function in 5 phases while respecting the 3-phase rhythm of the latter.

3.4.5. The US Literary System and the Bible

From a preemptive perspective, it seems important to discuss first the position of the Bible among the mixture of Indigenous and European influences that gave birth to the North American literature.

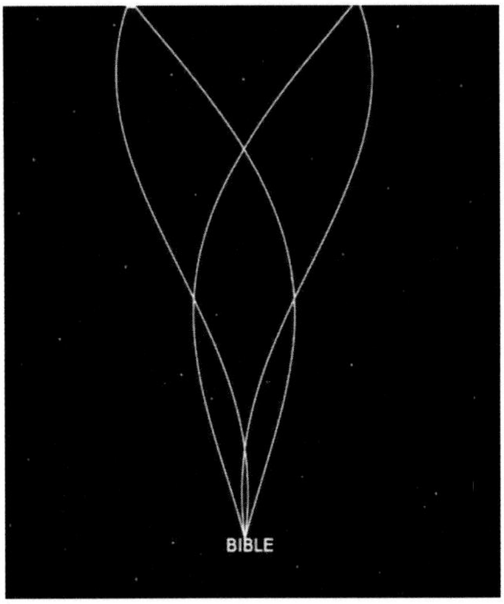

It can be safely assumed that the size of the literary system expanded since 1863 at the same logarithmic pace as the collection of the US Library of Congress (from +/- 200,000 in 1863 to 32 million printed books in 2020, averaging a yearly growth of 3.4 %).
North American literature produces yearly about 1,000,000 new fiction and non-fiction titles, totaling a sale of 675 million books. There are more than 168,000 Bibles that are sold or given to others in the United States every day, which averages to about 20 million yearly. For this reason, the Bible is excluded from book bestsellers lists because it would always be on top.
One of the most surprising and most profound insights from the science of memetics is that your thoughts are not always your own original ideas.

You catch thoughts—you get infected with them, both directly from other people and indirectly from viruses of the mind. People don‘t seem to like the idea that they aren't in control of their thoughts. It touches to the subject of the free will, which will become even a more contentious issue during the further development of this essay.
‘An egg has 'a hen's way of making another egg'.... An organism is a gene's way of making another gene ... *North American literature is the Bible's way of making more Bibles.*

3.4.6. Trend-lines in US literature.

3.4.6.1. Methodology

I started a tentative approach to achieve a mathematical model based upon the previous assumptions with an analysis of the bestseller book lists in the USA with the help of Fourier's theorem.

$$f(x) = a_0 + \sum_{n=1}^{\infty} \left(a_n \cos\frac{n\pi x}{L} + b_n \sin\frac{n\pi x}{L}\right)$$

If there are cycles in the bestseller lists of literary works, the low degree of the observed correlation might be due to the data of incomplete cycles; in the second place, the record is only drawn from the first book on the yearly bestselling list and might give an accidental, low degree of correlation between content and time. In the third place; the data for the period in American novels from before 1895 are mostly derived from the literary research of James D. Hart whose conclusions, as by himself admitted, were based on less solid historical records (for some years there is even not a given assumption of what might have been the most popular book).
The bestselling data for the nonfiction books are taken from a list drafted by Nathan Bransford. No relevant data could be retrieved for the period prior to 1917.
Before starting to analyze the literary data of the best seller lists, it was unknown whether there are many cycles or only one cycle or, indeed, whether there were any cycles at all. The author admits that the mathematical purists may have objections to the use of quantitative mathematical processes to calculate qualitative data, but the results indicate that it produces a workable transitional outcome indicating some cyclical trend-lines among the literary equilibriums.
The here proposed literary mathematical model as whole may be in need of some additional feedback by running it through a more powerful data processor than my laptop.

3.4.6.2. Trend-lines in US fiction between 1863 -2015.

In Figure 7 are the results detailed of a laborious examination of the annual themes in American bestselling novels, presented in graphic form by using their Dewey Decimal Classification (DDC) numbers. On the axis are measured, within assigned limits, the possible lengths of cycles in the 157 years of literary bestsellers.
The period before 1863 was dismissed for analytical data analysis because of its unreliable and incomplete nature that was also considered non-representative for a study of bestselling novels in America. We discover very soon that the period between 1863 and 2019 consists of two different subject cycles with a cleavage between 1964 and 1966.
From 1965 on the subject cycle of the novel is almost "flat-lining", with some feeble irregular oscillations between social and spiritual issues, suggesting that other information carriers have taken over some aspects of the social function of the printed novel.
The closest corresponding arithmetical cycle approximating the Nr. 1 US bestseller fiction subject data from 1863 till 1964 is represented by the equation (figure 2, marked in red);

$$f(x) = \sum_{i=13}^{113} \frac{1000}{180}\left(101.5 + 53.7\sin\frac{\pi x}{11} + 18\sin\frac{2\pi x}{7} + 13\sin\frac{\pi x}{13}\right)$$

It's not unlikely that the same fiction subject cycle from the period between 1863 -1963 still continues but includes nowadays a more diverse collection of information carriers. However, it's not to deny that the subject list of the literary fiction passed a bifurcation point in 1965. Consequently, 1965 has been ignored in the mathematically calculated trend-line and its value set on 0, which accidently nearly coincides with the subject DDC of the 1965 bestseller (archeology 07). The best approximate trend-line for the period 1966 – 2019, is based upon succeeding cycles of 18 years for which the formula is;

$$f(x) = \sum_{i=11}^{29} \frac{1000}{180}\left(650 - 1.3\tan\frac{\pi x}{18}\right) - \frac{1000}{180} abs\left(40\sin\frac{\pi x}{3}\right)\ldots$$

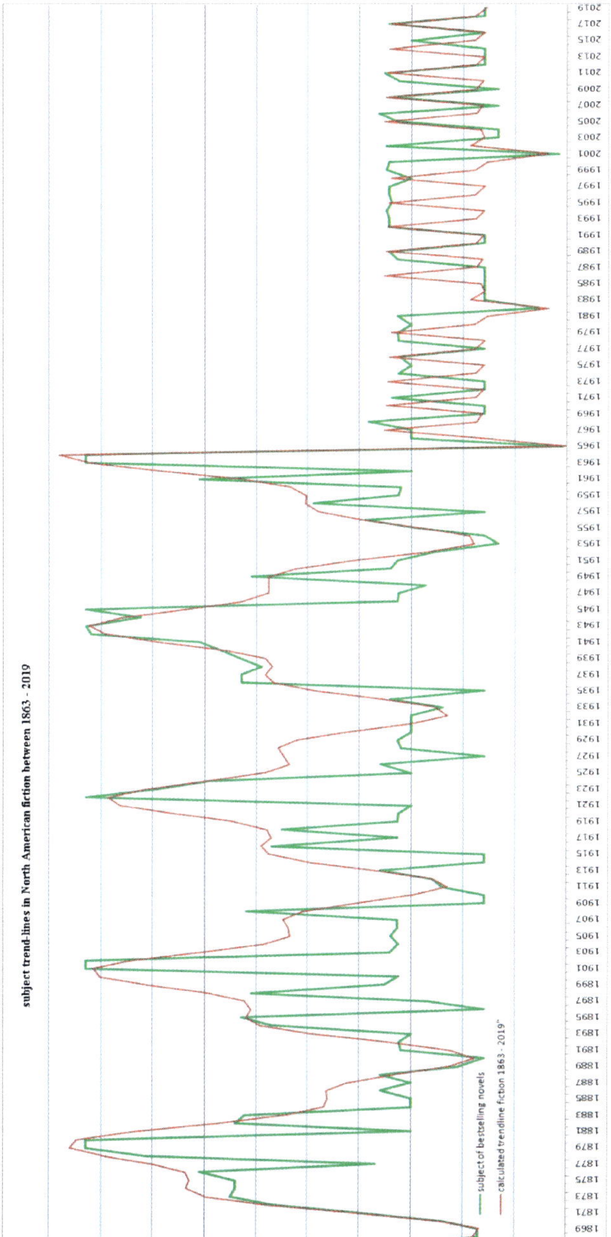

Figure 7: real and mathematical trend-line of subjects in the novels between 1863 - 2019

3.4.6.3. Protagonist fluctuation in American bestselling Novels between 1863 and 1963.

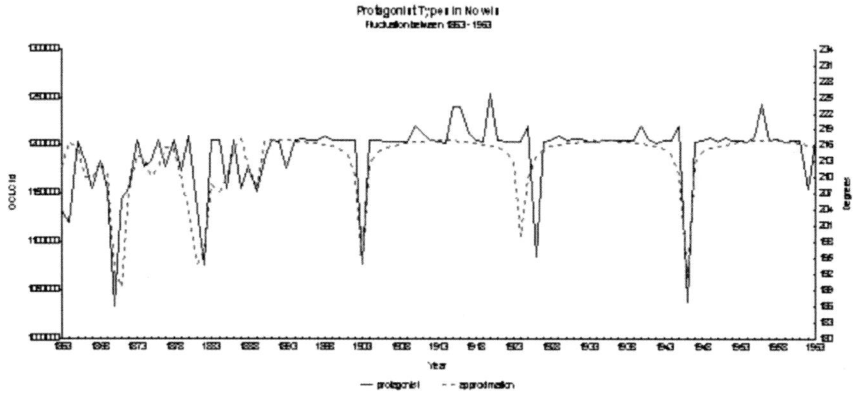

Figure 8: Table showing the fluctuation of main protagonists and its trend-line in novels over 100 years.

The types of main protagonists in the Nr. 1 bestselling novels 1863 – 1963 are classified by their OCLC number. We remark that during the period between 1863 and 1890 there is a bigger fluctuation of main protagonist types than in the following 73 years. I have no explanation for this phenomenon and can only conclude that for this timeframe the mathematical approximation consists of two synergetic equations;

For the period between 1863 – 1890;

$$f(x) = \sum_{i=1}^{27} \frac{1,300,000}{360} \left(214 - \text{abs}\left(2.95\tan\frac{\pi x}{11}\right) + 3.65\sin\frac{\pi x}{2.25}\right)$$

For the period between 1891 – 1963;

$$f(x) = \sum_{i=1}^{72} \frac{1,300,000}{360} \left(217 - \text{abs}\left(0.75\tan\frac{\pi x}{21.5}\right)\right)$$

I haven't been able to calculate an approximate for the protagonist trend-line after the bestselling fiction passed its bifurcation point in 1965. It's mostly flat-lining with some irregular outbursts.

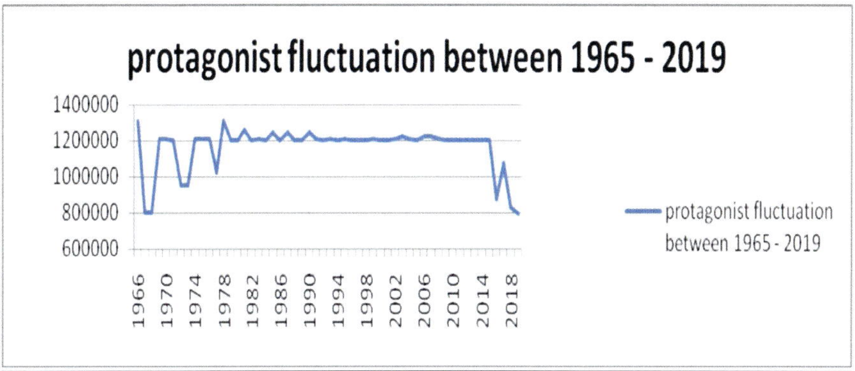

Figure 9: protagonist fluctuation 1965 - 2019

3.4.6.4. Subject periodogram of the American bestselling nonfiction between 1917 -2017.

After analysis of the records that were available during the composition of this essay, no apparent discontinuity in rhythm - or amplitude change can be detected over a period of 100 years. It appears that the subject cycle of the nonfiction literature is s ruled by a different set of mechanisms than those of the novel.

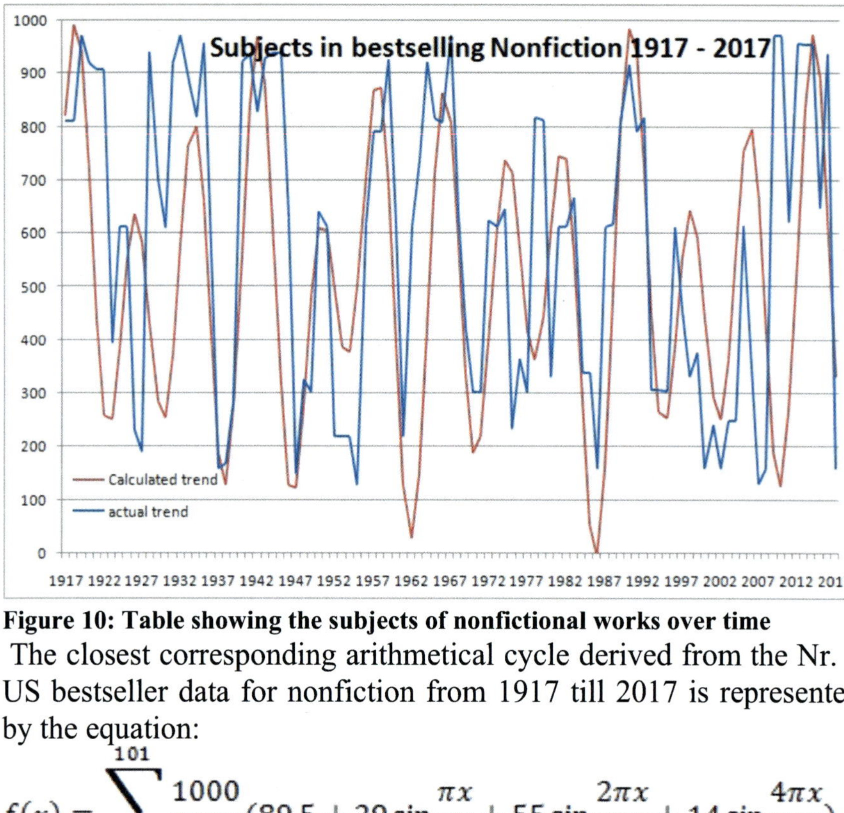

Figure 10: Table showing the subjects of nonfictional works over time

The closest corresponding arithmetical cycle derived from the Nr. 1 US bestseller data for nonfiction from 1917 till 2017 is represented by the equation:

$$f(x) = \sum_{i=1}^{101} \frac{1000}{180}\left(89.5 + 29\sin\frac{\pi x}{6} + 55\sin\frac{2\pi x}{8} + 14\sin\frac{4\pi x}{36}\right)$$

3.4.7. The Multivariate Mathematical Model of North American Literature.

After having constructed a simple algorithmic model, we can now proceed to a quantitative analysis of the literary system. This model can provide new insights into the structure of the system and increases our understanding of the system itself. Modeling can also reveal some logical errors of simple literary concepts.
For the sake of clarity and logical consistency, I've drawn the literary model as if it would be executed by a computer program, indicating from what stage on a human writer takes it usually over from the research algorithm. It details the process that produced the fiction and nonfiction in North American literature since 1863.

The literary mathematical model uses the collection of the US Library of Congress to build a semantic linked network through FCM (Fuzzy Cognitive Map). First, the system creates a list of keywords for the required story, then save it on a database and search any word from a keyword list. For this purpose, it uses a SLN (Semantic Link Network) to create an outline with its preferences, determined by the prevailing subject trend. I've indicated the place where a human writer usually takes it over from the research algorithm. When the outline of the story is available in plain text form, it is saved on another template and patched up to become a readable text.

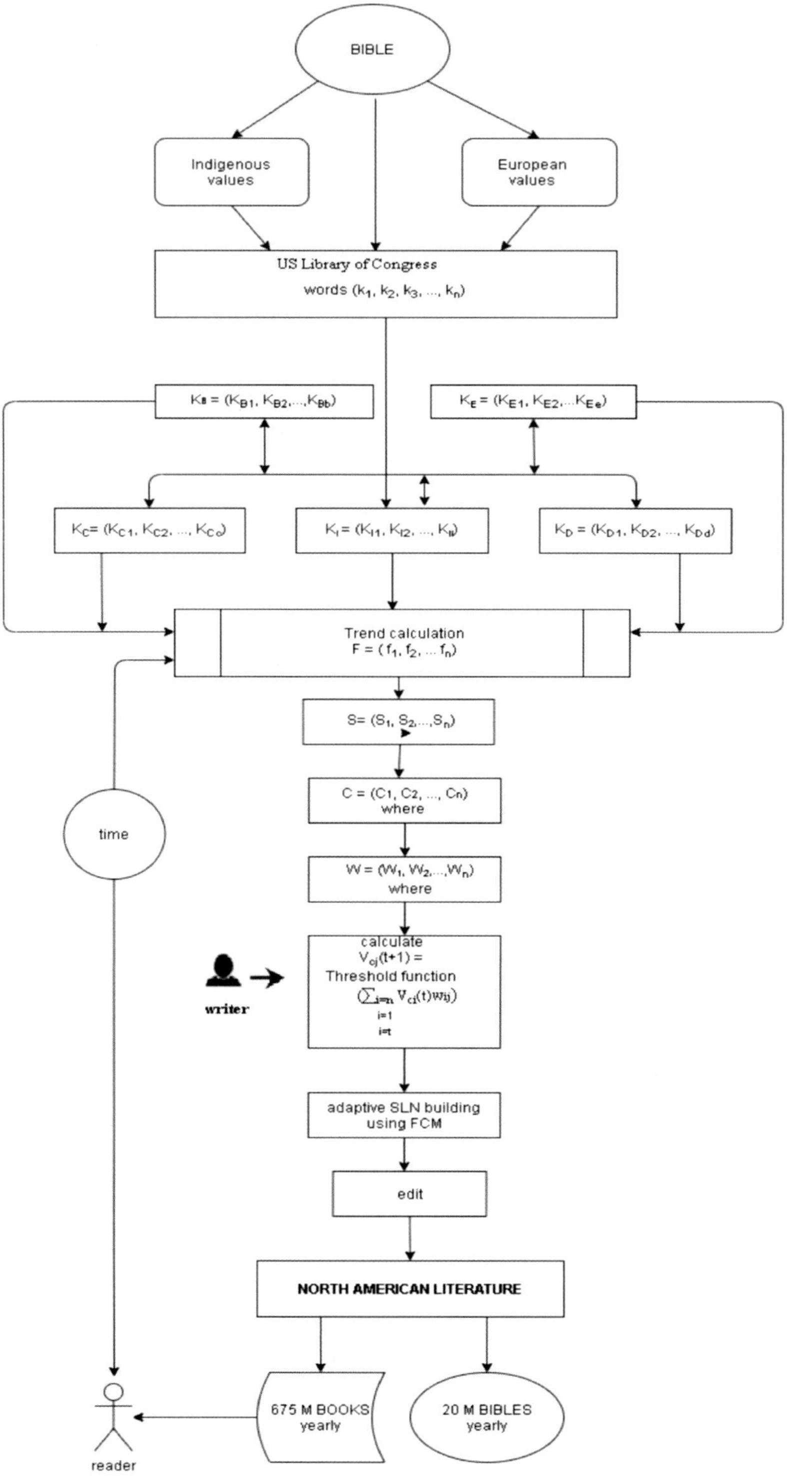
BIBLE
Indigenous values
European values
US Library of Congress
words (k1, k2, k3, ..., kn)
KB = (KB1, KB2,...,KBb)
KE = (KE1, KE2,...KEe)
KC= (KC1, KC2, ..., KCc)
KI = (KI1, KI2, ..., KIi)
KD = (KD1, KD2, ..., KDd)
Trend calculation
F = (f1, f2, ... fn)
S= (S1, S2,...,Sn)
C = (C1, C2, ..., Cn)
where
W = (W1, W2,...,Wn)
where
time
calculate
Vcj(t+1) =
Threshold function
(∑i=n Vci(t)wij)
i=1
i=t
writer
adaptive SLN building
using FCM
edit
NORTH AMERICAN LITERATURE
675 M BOOKS
yearly
20 M BIBLES
yearly
reader

K_B, K_E, K_I is the set of reserve words taken only from the Bible or European - and Indigenous values. K_C is the set of common reserve words; those are taken from the whole collection of the US library of Congress. K_D is the set of different reserve words that are not common in the collection of the US Library of Congress.
After defining/extracting reserve words, the system finds the frequency of those reserve words. Number of occurrences of a particular reserve words in a given database is called the frequency of that word, then it finds the state value of reserve word automatically.

First, the trend calculator makes a list of keywords for the required story, then save it on a database and search any word from a keyword list. IQSB built a SLN automatically and helps user to create a story with those preferences. Now the story is available in plain text form, which is saved on another database automatically, is ready for readers as one of the possible outcomes of related searches. Reserve words basically belong to a main story that the literary system wants to build. First, the system finds the words relevant in the Bible. Or combine relevant sources namely the US corpus (Bible, Indigenous -, and European values). Words are represented by $K = \{k_1, k_2. . ., k_n\}$. This set of reserve words is further divided into five subsets given as follows:

i. $K_B = \{k_{B1}, k_{B2} \ldots, k_{Bb}\}$
ii. $K_E = \{k_{E1}, k_{E2} \ldots, k_{Ee}\}$
iii. $K_I = \{k_{I1}, k_{I2} \ldots, k_{Ii}\}$
iv. $K_C = \{k_{C1}, k_{C2} \ldots, k_{Cc}\}$
v. $K_D = \{k_{D1}, k_{D2} \ldots, k_{Dd}\}$

Where K_B, K_E, K_I is the set of reserve words is taken only from the Bible, European values or Indigenous values respectively. K_C is the set of common reserve words; those are taken from this corpus.
K_D is the set of different reserve words; those are not common in the US corpus. After defining/extracting reserve words, find the frequency of those reserve words. Number of occurrences of a particular reserve words in a given database is called the frequency

of that word, then find the state value of reserve word automatically. Every reserve word has its own state value], $S = \{s_1, s_2, \ldots, s_n\}$ the meaning of state value is own worth of reserve word.

After finding this co-occurrence is required as completion of FCM [1,2] $C = \{C_1, C_2, \ldots, C_n\}$ where $C_i = \{C_{i1}, C_{i2}, \ldots, C_{in}\}$ and in a combined fashion, C can be represented as:

$$C = \begin{bmatrix} C_{11} & C_{12} & \ldots & C_{1n} \\ C_{21} & C_{22} & \ldots & C_{2n} \\ \vdots & \vdots & \vdots & \vdots \\ C_{n1} & C_{n2} & \ldots & C_{nn} \end{bmatrix} \quad (1)$$

weighting that co-occurrence of the reserve words required $W = \{W_1, W_2, \ldots, W_n\}$, where $W_j = \{w_{j1}, w_{j2}, \ldots, w_{jn}\}$, and in matrix form can be seen as:

$$W = \begin{bmatrix} w_{11} & w_{12} & \ldots & w_{1n} \\ w_{21} & w_{22} & \ldots & w_{2n} \\ \vdots & \vdots & \vdots & \vdots \\ w_{n1} & w_{n2} & \ldots & w_{nn} \end{bmatrix} \quad (2)$$

Now, all the material is ready for calculation, according to the FCM formula [1,2] for automatic building of an SLN with thrash-holding function:

Calculate $V_{c,j}(t + 1)$ = threshold_func

$$\left(\sum_{\substack{i=1 \\ i \neq j}}^{i=n} V_{ci}(t) w_{ij} \right) \quad (3)$$

Patches can be applied for building a strong story according to the writer's wish during the process. The system collects all related information from different resources by building a semantic network through fuzzy logic.

Now the system has collected all related information from different resources by building a semantic network through fuzzy logic for a concept for which the trend calculator needs a story. In the second step, a story is generated by the writer based on collected information. Proposed system is semi-automatic because first step is automatic, and second step involves a writer.

In this development, Fuzzy Cognitive Map (FCM) data mining techniques for extracting the data are implemented. The main source of this research is the US library of Congress, that is an inactive dataset and in plain-text form.
Following steps are carried out to build the proposed system. – Enlisting the concept (keywords) – Finding the frequency on element basis – Finding the frequency on section basis – Finding the frequency on document basis – Finding the concept value of each concept using the fuzzy formula – Finding the co-occurrence/s – Finding the weight of each link using Fuzzy formula – Building SLN from the library collection – Compilation of the results in the form of a story .

This essay choose not to take into account the evolution of the linguistically aspect of literature and the same goes for variations in writing style during the different episodes of American Literature; we will go out from the supposition that the time follows a linear pattern and that the language remained unchanged. Both forces are treated as a constant although in reality they have been the subject of fluctuations. This has been done to scale down the complexity of this mathematical model of literature.

It is important to realize that this literary model doesn't provide a qualitative description or an answer to questions about" how", whereas this quantitative description answers questions about" how much".

The mathematical model itself is the synthesis of the qualitative and quantitative assumptions.

The average length of a contemporary bestseller is 375 pages, set inside the USA and the protagonist is a female lawyer or detective. Romance is an upcoming segment: from 5 % of the sales in 1994 to 40 % on Amazon today. The average amount of words that a sentence contains is slowly declining: from 15 in 1863 to 10 in 2020 while the use of question marks increased from 6 to 11 over 1,000 words.

I heard people argue that Dickens was a bestseller, so were Balzac and many others. Commercial literature back then still had some thought-provoking thematic elements and those two novels in particular were more imaginative products. Sure enough, there was also plenty of pulp-fiction throughout history like those western dime-novels written last century.

The model's trend-line indicates that this year's subject for the bestselling fiction will most likely fall into the DDC interval between 000 – 060. Based upon past results this either indicates archeology, the antichrist or extraterrestrial life as a central theme for the 2020 bestselling novel.

Commercial fiction is meant to entertain and not to provoke thought or leave a particularly lasting impression while literary fiction is thematically deep, challenges conventional notions, leaves us with something to think about, has characters that aren't cardboard, and basically has aesthetic merit. Bestselling contemporary novels reflect the predisposition of the 21st century's homo ludens and their multitasking attention span.

Bibliography;

Herman Northrop Frye, Anatomy of Criticism: Four Essays (Princeton University Press, 1957)

Herman Northrop Frye. The Great Code: The Bible and Literature. New York: Harcourt Brace Jovanovich, 1983.

Jose Nilo, G. Binongo & M. W. A. Smith, A bridge between statistics and literature: The graphs of Oscar Wilde's literary genres, , pp. 781-787

Barry, P. (2002), 'Structuralism', Beginning theory: an introduction to literary and cultural theory, Manchester University Press, Manchester, pp. 39–60.

Selden, Raman / Widdowson, Peter / Brooker, Peter: A Reader's Guide to Contemporary Literary Theory Fifth Edition. Harlow: 2005. p. 76.

James D. Hart, The Popular Book, A history of America's literary taste, Oxford University Press, 1950

Robert Finkelstein, A MEMETICS COMPENDIUM, Robotic Technology Inc., University of Maryland University College BobF@RoboticTechnologyInc.com www.RoboticTechnologyInc.com May 2008

3.4.8. Analogies from scientifically Literature.

3.4.8.1. The propagation of literary works

This can be compared to the diffusion of large numbers of tiny particles and has been extensively studied[38], and is described by the equation;

$$\frac{\partial C(x,t)}{\partial t} = D\frac{\partial^2 C(x,t)}{\partial x^2},$$

where C(x,t) is the concentration of particles at location x at time t. This equation gives no indication as to the future motion of any individual particle. It is only a description of collective behavior.

3.4.8.2. The exploration of literary expansion processes described by differential equations.

When our data consist of observations y , at times t_i, $i = 1...n$ we can estimate the derivative between times t_i and t_{i+1} by $d_{i+1} = (y_{i+1} - y_i)/(t_{i+1} - t_i)$. By plotting d_{i+1} against $(y_{i+1} + y_i)/2$, an estimate of the average value of y, we can investigate the relationship between dy/dt and y. In the case of exponential expansion, we would expect the graph to show a straight line. Where growth is limited, we would expect to see some departure from linearity. For example, volume growth of printed books, described by the equation

$$\frac{dy}{dt} = ry(a-y),$$

---

[38] Abramowitz, M. and Stegun, I.A. 1968. Handbook of Mathematical functions, Dover, New York.
Press, W.H., Flannery, B.P., Teukolsky, S.A. and Vetterling, W.T. 1987. Numerical Recipes, University Press, Cambridge, UK.

should lead to a quadratic curve. Since it is generally rather difficult to determine visually whether or not a curve is quadratic, we would like to find a plot which gives a straight line when growth follows the expansion equation. To motivate this plot, we write the expansion equation as

$$\frac{1}{y}\frac{dy}{dt} = ry(a-y)\,,$$

If we could calculate the left-hand side, sometimes called the proportional expansion rate, from the data, then plotting against y would indeed yield a straight line. There is a mathematical result which says that

$$\frac{1}{y}\frac{dy}{dt} = \frac{d}{dt}(\log y)\,,$$

so if we calculate $(d_i = \log y_{i+1} - \log y_i)/(t_{i+1} - t_i)$ we can proceed as above. This is illustrated in Figure 4. Deviation from linearity in the fourth plot not only indicates that the expansion is not the correct model for the data, but it may also suggest the kind of adaptation, which is appropriate.

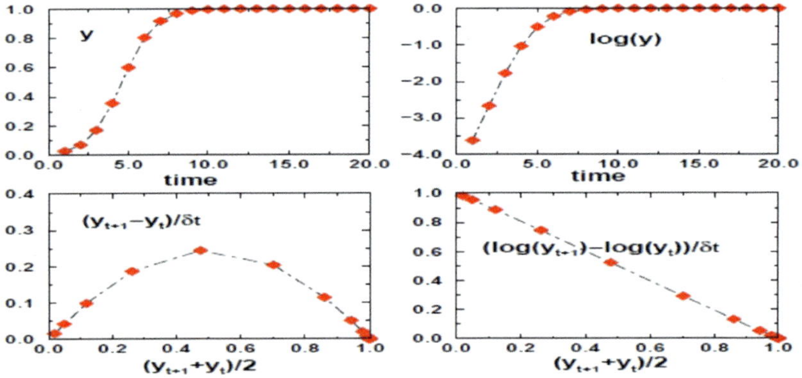

We observe that a complete analytical solution for a stochastic model involves finding the distribution of outcomes but are contended if we can solve the equations for the mean and standard deviation. In common, getting an analytical solution for a literary problem is rarely a simple matter.

3.4.8.3. The use of a conical coordinate system.

A coordinate system is defined by a series of vectors and an origin point. This is a useful simplification of the possibilities; this is true for any coordinate system that follows the rules of Euclidean geometry. Conical coordinates are non-Euclidean. Where in Euclidean geometry, the sum of the angles of any triangle should add up to 180 degrees exactly is not true for conical geometries or conical coordinates. This is because "lines" in conical geometries are curves when seen relative to Euclidean geometries.

An additional advantage is that conical coordinates are three dimensional and have 3 values. One value, given the name "r" (for radius) of the respective concentric spheres and by two groups of perpendicular cones that are respectively aligned along the z- and x-axes.

This is much easier to see in diagram form:

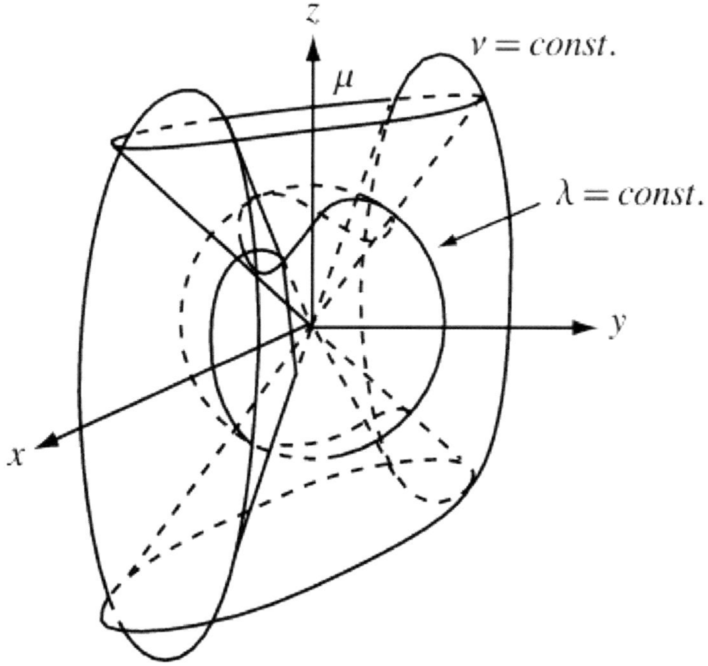

Figure 5 Conical Coordinates[39]

The feature that makes a curvilinear coordinate system attractive for the expression of a mathematical model of literature is that it allows us to position an object around another object, especially if it requires moving along spheres relative to another object.

The conical coordinates (λ, μ, ν) as defined by Wolfram Language are:

$$x = \frac{\lambda\mu\nu}{a\,b} \quad (1)$$

$$y = \frac{\lambda}{a}\sqrt{\frac{(\mu^2 - a^2)(\nu^2 - a^2)}{a^2 - b^2}} \quad (2)$$

$$z = \frac{\lambda}{b}\sqrt{\frac{(\mu^2 - b^2)(\nu^2 - b^2)}{b^2 - a^2}}, \quad (3)$$

where $b^2 > \mu^2 > c^2 > \nu^2$.

Byerly used a similar (r, μ, ν) system but replaced λ by r, a by b, and b by c.

The preceding equations give us

$$x^2 + y^2 + z^2 = \lambda^2 \quad (4)$$

$$\frac{x^2}{\mu^2} + \frac{y^2}{\mu^2 - a^2} + \frac{z^2}{\mu^2 - b^2} = 0 \quad (5)$$

$$\frac{x^2}{\nu^2} + \frac{y^2}{\nu^2 - a^2} + \frac{z^2}{\nu^2 - b^2} = 0. \quad (6)$$

For the scale factors we calculate;

$$h_\lambda = 1 \quad (7)$$

[39] Weisstein, Eric W. "Conical Coordinates." From MathWorld--A Wolfram Web Resource. http://mathworld.wolfram.com/ConicalCoordinates.html

$$h_\mu = \sqrt{\frac{\lambda^2\left(\mu^2 - \nu^2\right)}{\left(\mu^2 - a^2\right)\left(b^2 - \mu^2\right)}} \qquad (8)$$

$$h_\nu = \sqrt{\frac{\lambda^2\left(\mu^2 - \nu^2\right)}{\left(\nu^2 - a^2\right)\left(\nu^2 - b^2\right)}}\ .$$

Not all positive definite metric forms relate to Euclidean space. When there is not a coordinate transformation which reduces a given metric form to that of

$$dl^2 = (dx^1)^2 + (dx^2)^2 + (dx^3)^2$$

indicates that this metric form is not Euclidean. This remarkable mathematical result is utilized in the General Relativity Theory.

The background of the mathematical model at hand expands with a yearly average coefficient of 3.42 %. The earliest record[40] available dates from 1866 and shows an inventory of 183,142 books. By 2014, this amount was inflated to 31,127,906 books.

---

[40] Annual report of the Librarian of Congress. By: Library of Congress, Published: (1866)

3.4.8.4. Improving the validation of the obtained results.

It was suggested to add a Bayesian validation function to the synthetic component of the model. The principal technique used by this function is an interval-based hypothesis testing, conducted on the reduced difference data to assess the model validity under uncertainty.

Cowles et al. (2009) elaborated that *"...in the Bayesian paradigm, current knowledge about the model parameters is expressed by placing a probability distribution on the parameters, called the "prior distribution", often written as*

$$p(\theta).p(\theta).$$

When new data ***y*** *become available, the information they contain regarding the model parameters is expressed in the "likelihood," which is proportional to the distribution of the observed data given the model parameters, written as*

$$p(y/\theta).p(y/\theta).$$

This information is then combined with the prior to produce an updated probability distribution called the "posterior distribution," on which all Bayesian inference is based. Bayes' Theorem, an elementary identity in probability theory, states how the update is done mathematically: the posterior is proportional to the prior times the likelihood, or more precisely,

$$p(\theta/y) = \frac{p(\theta) \times p(y/\theta)}{\int \Theta p(\theta) \times p(y/\theta) d\theta}.$$

In theory, the posterior distribution is always available, but in realistically complex models, the required analytic computations often are intractable. Over several years, in the late 1980s and early

1990s, it was realized that methods for drawing samples *from the posterior distribution could be very widely applicable*" (cit.)[41].

Finally, it was decided not to add a Bayesian function to the model because it is a complication that surpasses the scope of this essay. The best way to improve the working of the model is to add additional bibliographic data and to tune up the proposed algorithms.

[41] Kate Cowles, Rob Kass and Tony O'Hagan, What is Bayesian Analysis? International Society for Bayesian Analysis, 2009.

3.4.9. The use of the model to forecast literary trends

Earlier on in this essay it has been stated that the proposed model has a quantitative concept. It does not give any qualitative judgment about the literary works but gives an indication of some of their bestselling aspects.

In theory it should be possible to forecast future trends with the proposed mathematical model, but it would only work with a relative accuracy for the non-fiction literature since some functions of the literary fiction have been usurped by other media. One might sometimes get funny answers.

The problem lies with the solidness of the system caused by small disturbances from outside. For example, neighboring equilibrates can fusion and next disappear completely; the solutions that were converging from this equilibrates, are obliged to move to somewhere else, what causes a drastically change of behavior on the long term. Because in this case it is only a stochastically disturbance, is there also uncertainty about the moment that the system passes such a bifurcation point. Still the possibility exits of disturbances in the system occur due to our incapacity to create exact conditions at the beginning of the process[42].

[42] Ruelle, David; Takens, Floris (1971). "On the nature of turbulence". Communications in Mathematical Physics 20 (3): 167–192.

3.4.10. Indications of unexplored fields in US literary criticism.

The whole idea of the possibility to create a forecasting mathematical model of literature is putting fundamental questions to the nature of this field, its origins and significance for society that we don't want to discuss into this essay. We just satisfy ourselves by observing that literary history, - contend and – development are governed by a law of nature that pursues its own purpose.

Also, the introduction of new technologies is influencing the ways that trends and ideas are propagated and should retain the attention of the literary community. We refer especially to the abrupt change of the subject cycle in novels in 1965.

The phenomenon that the non-fiction subject cycle remained unaffected by this sudden change indicates that this fraction of literature is governed by different mechanisms and deserves a closer look than it received up to now, with the tendency of literary criticism to focus on fictional literature or non-fiction works in its own field of experience and interest.

A minor issue of academically interest is the question why the protagonist function in novels fluctuated much more in the period 1863 -1890 than in the period 1891 -1964 and is almost flat lining since then.

The model also doesn't consider the e-books, mainly sold by Amazon, another field that needs a closer look. Several thousands of newly publications hit the market every week through this channel and their relative success depends mostly on the marketing skills of their author. Some undiscovered literary pearls must be lying rotting there.

An adaptation of the proposed literary model to the system theory would greatly improve its forecasting function and enlarge our insights into the functioning of the literary system as a whole, but was also omitted in this essay that only wants to introduce a basic methodology for the creation of a mathematical literary model.

Almost all lanes of a dynamic process are converging to a limited partition; the attractor. Such a system is called a dissipative system. The movements in such a system are predictable, while small differences of the beginning values are causing qualitative equal solutions. By chaotically systems is the attractor a complex figure and by enlargement of a part of this figure an equally complex figure can be found back. In such a case, it is called a strange attractor[43].

Strange attractors are geometrical objects that by random enlargement show increasing more detail. It is possible to create complex structures with simple rules.

Visual representation of a strange attractor

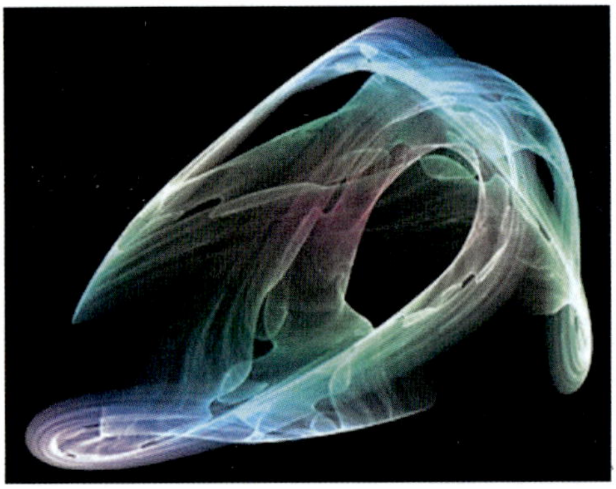

Fractals are used for the design of theoretical models of crystals growth and the coding and decoding of cloud structures. Chaotically systems have fractals as attractor. Fractals are functioning as separation between different attractor areas. This means that for solutions that are starting in the neighbor of the fractal separation line, no possibility exists for a unanimous prediction for the long term.

An example of such fractal is the Mandelbrot set[44] with following sequence

$(c, c^2 + c, (c^2+c)^2 + c, ((c^2+c)^2+c)^2 + c, (((c^2+c)^2+c)^2+c)^2 + c, ...)$

[43] Gouyet, Jean-François (1996). Physics and fractal structures. Paris/New York: Masson Springer. ISBN 978-0-387-94153-0.
[44] Tan, Lei. The Mandelbrot set, theme and variations. Cambridge University Press, 2000. ISBN 978-0-521-77476-5. Section 2.1, "Yoccoz para-puzzles", p. 121

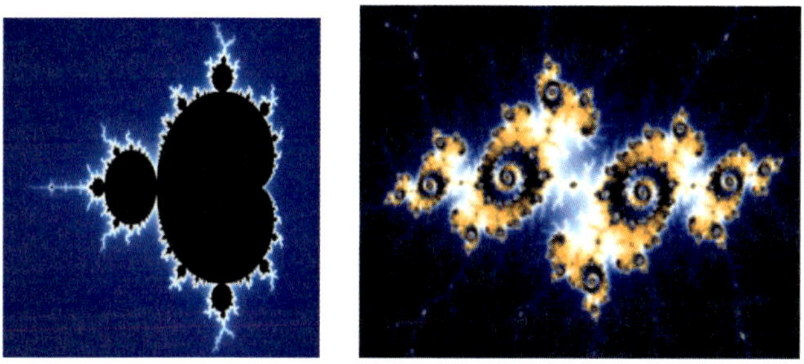

Initial image of a Mandelbrot set zoom sequence with a continuously colored environment. The Mandelbrot set has become popular outside mathematics both for its aesthetic appeal and as an Example of a complex structure arising from the application of simple rules, and is one of the best-known examples of mathematical visualization.

Conservative systems are ideal, closed systems, where the evolution is contained. Because of the conservation of energy, such systems can be described by a curve on a so-called energy surface. The conclusion was that in even such systems there was no possibility to exclude chaos on long term because of change of the parameters.

The transition from order to chaos shows several universal aspects. When on a horizontal axis a parameter is marked and on a vertical line de corresponding attractor, is there a clear observation of a number of sequential period doublings till the attractor becomes chaotically. The parameter interval where in a certain period occurs is by each period doubling shortened with a factor with limit d=4,669. Finally, this period doublings result into chaos.

The universality of the Feigenbaum scenario[45] leads to the expectation that for a number of global systems the specifications of the components are irrelevant.

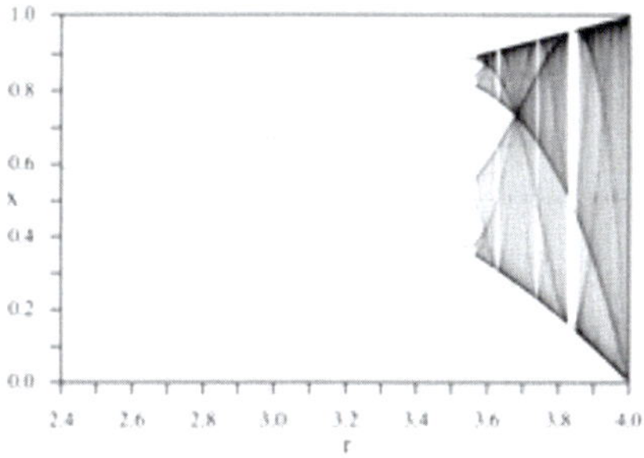

Bifurcation diagram of the logistic map. Using fractal geometry to describe natural forms such as coastlines, mathematical physicist Mitchell Feigenbaum developed software capable of reconfiguring

[45] Feigenbaum, M. J. (1978). "Quantitative Universality for a Class of Non-Linear Transformations". J. Stat. Phys. 19: 25–52

coastlines, borders, and mountain ranges to fit a multitude of map scales and projections. Dr. Feigenbaum also created a new computerized type placement program which places thousands of map labels in minutes, a task that previously required days of tedious labor.

3.5. What about Awareness?

3.5.1. Definition of Awareness.

The amount that a creature is conscious depends upon the logical structure of the algorithm that evokes the mental condition. The physical incorporation of the algorithm is irrelevant because the algorithm has a kind of immaterial existence that is not related to its realization in physical objects. The activation of the algorithm invokes a condition of consciousness.

Intelligent behavior is sometimes defined as the behavior that is adaptable variable during the span of lifetime of an individual and has three fundamental characteristics;

1. The capacity to react in an unusual way towards the stimulus-situation.

2. The ability to remember learned lessons.

3. Having the capacity for generalization.

There is no reason to believe that the self-conscious part of noosphere has to be limited to carbon dioxide-based creatures. Following this concept is awareness in fact a program that runs upon the hardware of our body and brain. Already the internet is a vast part of the noosphere, and it is unpredictable if someone will ever develop some algorithm that will give it self-consciousness. Further mutations of human mind and body, giving humans more abilities, are not to exclude. Bioengineers are tinkering at the DNA of our food; it's just a question of time before they will start doing the same with the human genome. Meanwhile, the quest for a better understanding of the human mind and body is still an ongoing story.

3.5.2. Definition of a Person.

A person is a complex system, made up of component subsystems. Person is, in principle, the entire self-organizing, multilevel, causal thicket, including bodily, mental, and social aspects, and representations of past and future organizations (selves). This list is meant to be open-ended, and other dimensions can be added as needed. Person, of course, is embedded in a larger complex environment (the universe). My new includes all of these, always. It contrasts with common usage, in which just how much is included by person and personal always depends on the speakers' conventions and on their purposes. Thus, under the common definitions, sometimes a person refers to the body ("touched his person), sometimes to intimate feelings ("religion is too personal"), and sometimes to social relations ("separate personal and professional life"). Since this is not made explicit in ordinary discourse, it contributes to the confusion.

Our common speech sometimes distinguishes between person and self, but with multiple, context-dependent senses, … I propose that we formally designate that person is extended over time, and self is the current organization of the person. That is, self is the way all of the subsystems of the person are related to each other. I emphasize that self is a characteristic of the person as a whole, rather than just another subsystem or constituent as some psychological models would have it. In discussing the complex, we call person it is very useful to be able to refer to its organization (self) in contrast to its instantiation (embodiment). Organization is simply the set of all the relations among the constituents of the system. This can include such relations as membership, connection, and control.

It is readily apparent that a person, like any complex system, might be capable of several different patterns of organization (selves). For example, it may be tightly organized, with all subsystems integrated, so that the activity of each part is always affected by the state of all of the parts. Another system may be more loosely integrated so that some subsystems may function semi autonomously, within broad limits. A third pattern of organization might have some subsystems tightly integrated, acting like clusters, but only loosely coupled to other subsystems or clusters. This sort of description applies, no matter what sort of subsystems one is concerned with. A cognitive psychologist for example, might be concerned with the connections and control relations between the memory subsystem and the emotion subsystem; under some organizations memory may function largely independent of mood, and under other organizations might be greatly affected. To take an example from social personality theory, occupational behaviors may or may not be affected by drastic changes in family life (marriage, divorce). Defining the self as the pattern of organization emphasizes that it changes over time; over a period of years, as in the maturation of adolescence, or over minutes, as in multiple personality disorders. By defining self as the organization of subsystems rather than as just another subsystem, we get a clear referent for common phrases such as "a more integrated self," or "a development of the self," or "a loss of self." The concept of self as organization, varying dynamically in degree and quality, works at the neurological level of description as well as at the psychological level, and across levels .[46]

[46] *The concept "self" and "person" in Buddhism and in western psychology*, 4a. PERSON AND SELF ". NY:Columbia University Press. 2001.

3.5.3. The Organization of the Human Psyche

The theoretical possibility that the algorithm that invokes consciousness[47] is using quantum effects deserves also closer research. In this context one can argue that human irrationality is not simply a regrettable freak of nature, but a condition as necessary as rationalism.

While the mind knows a material preconditioning, it has also a will. This free will supersedes his material origins but stays anchored into the stream of the creation and its laws. There exist a multitude of psychological mathematical models, but the synthetic nature of this essay favors a simplified conceptual approach of this complex issue.

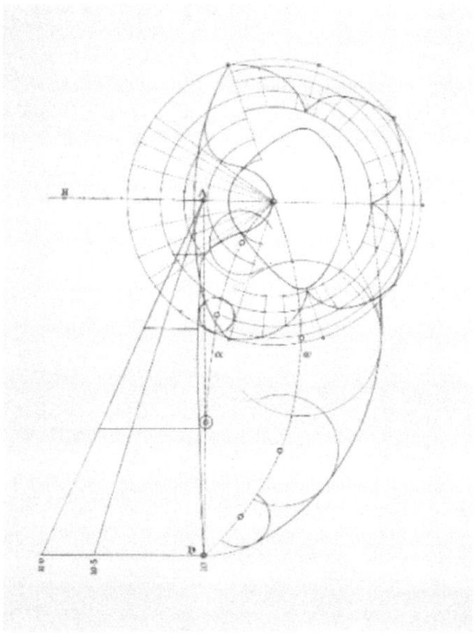

The working of the human psyche can also be compared with a geometrical process whose structure is determined by environmental and biological co-variation. It is established that all relevant personality data are stored in seven different clusters. The E(motional), I(ntelligence) and G(ender) clusters are generated by the action of the three hemispheric dichotomies that are superimposed on it while the other four are simple bipolar functions. R represents the superego or degree of conscientiousness; T represents the axis between practical and theoretical attitude; C represents someone's social attitudes or conventionality (progressive or conservative, left or right); A is the measure of agreeableness or the characteristics that define a person as sympathetic or tending toward dark triad behavior (Narcism, Machiavellism, and Psychopathy). These qualitative parameters are quantified and incorporated in basic concepts of algebraic topology.

[47] L. S. Cook, *Geometrical psychology, or, The science of representation: an abstract of the theories and diagrams of B. W. Betts*, 1887.

To get a comprehensible two-dimensional representation, the seven axes of the model have to be turned till one axis runs through all other clusters.

3.6. Will the Future Belong to the Homo Economicus or to the Homo Informaticus?

3.6.1. The current economic model.

In our actual model of society, there are four ways a person can provide for himself [48];

1. Getting paid for doing a job. In most Western countries you'll just earn enough to cover for your basic needs; food, clothing, shelter … In some Third World countries even not that.

2. Getting a bigger pay for doing a job that needs some specialized skills. Medical Doctors, Lawyers, Accountants et al. They earn good money but have to work menial hours to pay off their student's loan. Very often they feel also that society requires them to keep up appearances and put therefore a lot of their income on keeping up status; Big house, car, education for the children... Therefore, their lifestyle is on the limit of what their budget can afford and very often they work way over their 70th birthday before they can afford to retire while upholding their way of life.

3. Making money by letting other people work for you. Here you find the entrepreneurs and the CEO's of big companies. They created or oversee a big company and earn big. Just as the previous category, they work menial hours and are very often away from home on business trips. They also feel compelled to uphold status; membership of a golf or country club, a trophy wife, luxury cars, big villa. As a consequence, they have very often a volatile family life and early dead toll caused by stress related diseases.

4. Let your money work for you. These are the people who accumulated a minimum wealth of 1 million USD. Conservatively invested, this would give you a yearly income, after tax and costs deducted of 5 % or 4,100 USD monthly. For most people this is more than enough to cover their needs. When on top of that you decide to go to live in a third world country, you can put half of it aside and so even make your wealth grow. These people are called

---

[48] Andreas Eschbach. Eine Billion Dollar, Lübbe 2001

financially independent, since they don't have to work anymore to earn an income; their money does it for them.

The dominating idea is that one should procure as much money as possible or, for the smarter ones, as much as you need, to barter that for goods and services that you cannot provide by yourself. The ones, who earn not enough money to cover for their basic needs, are called poor. Those who earn more money than they need could be called rich. The latter condition is generally preferred and can be reached by two ways; either to increase your income till it covers your needs or by reducing your needs of acquisition till they're surpassed by your income. Those people are defined by Alvin Toffler as prosumers [49] .

For that purpose, people have created an economical system that is governed by the eight rules for warfare. They are;

1. TARGET; make money to cover at least for my needs

2. CONCENTRATION; from whom, where, how and when. Make sure that the target you are aiming at wants your product or service and can pay the price that you're asking.

3. ATTACK; create or indicate the need. Indicate your uniqueness in case of eventual competition.

4. MOBILITY; be on the internet and all other media into your reach. Create effective distribution channels. Have outlets where you can sell and eventually buy the stuff, service or information you need to be ahead of everyone else.

5. SURPRISE; be inventive. Go to the market with a new approach or product.

6. SECURITY; be careful what you disclose that is not protected by Law or that can make you vulnerable for competition. Don't do business with lovers, friends or relatives if you don't want to lose them over a money issue and the risk that the integrity of your operations' security being breached and consequently made vulnerable for hostile attacks.

---

[49] Toffler, Alvin (1970). The Third Wave (1980) Bantam Books ISBN 0-553-24698-4.

7. EFFECTIVITY; get feedback of used methods from independent sources; accountants, marketers, consultants, clients, etc....

8. THE USE OF FORCES; evaluate the flop potential and keep in mind that in some situations cash is King. A lot of promising people descended into poverty and personal bankruptcy because of neglecting those adagios.

Therefore, the end of the 20th century and the first decade of the 21st century was characterized by small scale civil wars, coinciding with returning waves of terror, drugs dealing and environmental sabotage. They are so called first wave wars for agricultural oriented cultures, but technically highly developed countries, on their way towards an on knowledge-based economy, got involved against their own will[50].

The US government has a public debt of 400 trillion on which they must pay 3 Trillion USD on interest rates, of which they must borrow one trillion [51] and so adding to the debt. The Federal Reserve has been buying for 2000 Trillion on bonds from the banks by simply crediting their accounts.[52] The mistakes of 1927 have been avoided and the result of the implosion of the real estate market was not the start of Great Recession II but of a regular recession.

The only problem is that nobody knows how to proceed further. When Bernanke, into his last weeks as chairman of the Federal Reserve, floated the idea of reducing the stockpile of bonds that the Federal Reserve owns, the stock markets went skydiving, so the Federal Reserve organized fast a press conference to say that they had no intention to do so into the foreseeable future. But these 2000 Trillion on bonds are still a ticking time bomb and when nobody comes up with some brilliant idea of how to dismantle it, it is for sure going to explode. Not to mention a government that must borrow money to pay one third of the interest rates of their debt. What will happen when the artificial low interest rates go up again? Right now, interest payments constitute 6% of the federal budget. A

---

[50] Toffler, Alvin, Powershift: Knowledge, Wealth and Violence at the Edge of the 21st Century (1990) Bantam Books ISBN 0-553-29215-3.

[51] "Federal debt basics – How large is the federal debt?". Government Accountability Office. Retrieved April 2014.

[52] Fannie Mae, Freddie Mac to Be Kept Off Budget, White House Says (September 12, 2008), bloomberg.com

devaluation of the dollar is looming as the only alternative for the next administration. And that could for sure trigger a Great Recession[53].

Consumer confidence is low, joblessness rates are high and police officers are waging an uphill battle against the street gangs and terrorists. The gangs are terrorizing whole neighborhoods with their territorial fights for drugs distribution; the mood on the street is sour. People lost their confidence into the political elite of their country who're pursuing personal goals rather than the interest of the population they pretend to represent. The sale of guns is sky rocking and so do public shootings. People are increasingly caving in for an implosion of the system and congress is dead locked by bipartisan issues. When politicians are indisposed to govern a country by incapacity of working out compromises, the socio-economic system will, by its own inherent structures and imperatives, impose one. The USA is clearly over its peak and slides into what the SETI mathematical evolution model calls the "senility time", while China is in its adolescent phase and moves towards a peak.

There also flows more energy into the agriculture than there exists in the form of basic materials. T.M. Malthus (1798) argued that a human population tends to outstrip available food supply because the population grows in geometric progression while available arable land grows only in arithmetic progression[54]. Into our current economic model, too much value gets wasted during the distribution process. This is especially true in third world countries that suffer of overpopulation, starvation and bad infrastructures where HALF of the food is wasted before it reaches the market. There must be quota set for the population density that a certain region can contend, in function of its capacity to feed them. Big monocultures must be replaced by more diverse agricultural production techniques. This throw away economy turns upon the underpaid materials out of the

---

[53] Greenspan, Alan (June 18, 2010). "U.S. Debt and the Greece analogy". Opinion Journal [online]. Retrieved February 2015

[54] The Malthusian and the logistic models of population growth represent the simplest use of classical mathematical methods in demography, the branch of sociology concerned with the size and composition of human populations.

third world and structural overproduction that makes the world markets smaller and smaller.

On top of that, a religious- fascist derivate of the Islam can lead to the shielding off from the oil sources. This will lead to a rise of the gold prices, a sharp rise of the inflation rate, a high joblessness among the active population, shortages and riots.

Therefore, is there an urgent need for a dissident economy apart from the current agro-industrial complex. In every economy, nature is the most important workforce and should also be valued as such. Environmental exploitation is the unwillingness to calculate into the cost the labor of the living nature that has to be able to work on a durable way, delivering raw materials. Modern technology gives humans the ability to do more work by themselves by giving local communities the possibility to provide by themselves for the most basic products, services and utilities. So is for the feeding of one human a surface of 18 is needed, a fact that communities in third world countries who're striving for a higher degree of autarky should keep in mind in view of the earlier in this chapter mentioned food distribution problems, which in this particular case extend to all facets of centralized services and products.

The internet is the most important tool that modern technology has provided us with since the invention of the book press. Right now, for most people, it is just a tool for amusement and superficial communication. Where the invention of the printed book caused the renaissance and the 17th century's illumination movement, the internet is for most people just an additional distribution channel for fun, gossip and movies. It can be much more. It has the potential to become the engine of a worldwide shift into economics, education, politics, religion, science, art and social behavior. It has already shown its capacity for new and popular ideas and trends to go viral. In the Arab world it has been instrumental for the spreading of the Arab Spring a couple of years ago. Just a pity that it was not better managed, because very soon the organizations claiming the popular upraise, proved to be nothing better than the ones they replaced. In Syria, for example, it created a power vacuum where ISIS installed a terrorist caliphate; the cure ended up to be worse than the disease.

Also here is decentralization of the production capacity and a turn towards what a community can produce out of its own renewable resources the next logical way. But of course, this idea runs into resistance by the existing energy providers, whose wealth and influence depends upon a centralized energy production.

New technologies and media can become the motor of an ecological future with flexible, small scaled structures and networks that are generating by themselves a post-materialistic culture. Next to product minded thinking there arises process minded thinking. Holism is integrating itself in all structures of knowledge. A new religion should help people to recognize and approach everything as a whole. The better people understand what moves things around us, the better they can position themselves to take advantage of these movements for personal growth and, subsequently, act in a way that makes the world a better place to live in for everyone.

3.6.2. Gross National Happiness is more important than Gross Domestically Product.

Throughout modern history, societal progress has been measured in terms of GDP. The higher the GDP, the more developed a country, so goes the general belief. There is, thus a mad rush for increasing GDP.

Hence, over time, GDP came to be seen as a surrogate for societal wellbeing – something it was never designed to be.

If the basic purpose of development were changed from the pursuit of profit to the pursuit of higher wellbeing in all its dimensions, the true level of happiness on the planet would certainly go up. The Stiglitz Sen Fitoussi Commission stated this. The Beyond GDP initiative in Europe recognizes it.

For Bhutan, the drive to glimpse the true nature of development began early under the leadership of their kings. Bhutan aspired to become a country where progress was holistic, inclusive, equitable and sustainable. Where political and spiritual matters were in balance. The four pillars of the Bhutanese model are; Good Governance, Sustainable Socio-economic Development, Preservation and Promotion of Culture and Environmental Conservation.

Over forty years since its introduction, Bhutan has ensured that GNH has been the unifying force behind all policy formulation and has shaped the country's five-year planning cycle. From a GNH perspective, it is understood that a decline in traditional heritage and cultural wisdom will lead to a general weakening of society. If you look around, many countries have lost much of their culture with the dynamic changing times. Bhutan, despite her lack of military might or economic power, maintains a distinct authority and a special identity of her own.

They adopted innovative technologies to grow new green sectors in the economy as a foundation for the creation of new and decent jobs in the economy. With the increase in urbanization and a growing middle class in Bhutan, they hope that the demand for modern transportation to support its urban lifestyles can be met by the carbon neutral electric car option. Within the next years, Bhutan will move to consolidate their sustainable and organic agricultural practices as even more reinforced policy.

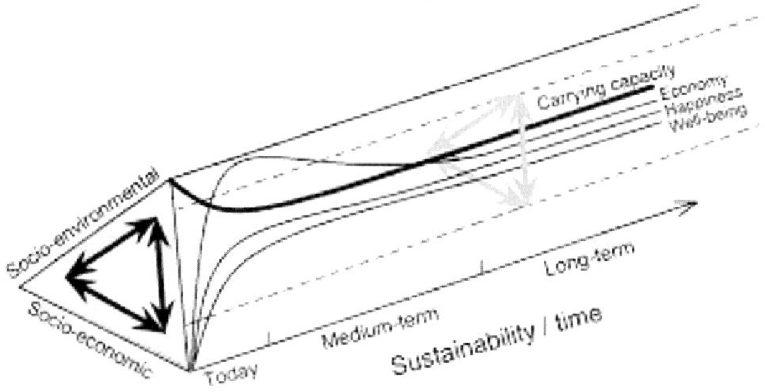

The current business model, with its over emphasis on profit maximization, on increasing shareholder value at almost any cost to environment and to the community, is unsustainable. Businesses must explore fostering happiness and wellbeing as an alternative business purpose. Bhutan will look inwards towards developing and empowering their small businesses, creating a climate that supports small business growth and encouraging citizens and young graduates to create businesses rather than seek employment. By investing in the growth and development of small businesses across the country, the massive result to be realized from this effort can facilitate economic self-reliance for Bhutan and build GNH business culture from the ground up.

Overall, in Bhutan, there was a significant increase in the subjective happiness in 2015 as compared to 2010. Subjective happiness has fascinated many researchers because it adds new quantitative data. But it does not capture altruism or responsibility. It does not reflect care for the environment either – the happiest countries worldwide include those that are polluting our planet most. So it is not appropriate as a standalone goal for society.

Looking at people's satisfaction with their health, it decreased – but their objective health improved. Turning to immediate family relationships, satisfaction with these improved but the GNH indicators worsened by a small amount but it is statistically significant. And subjective satisfaction with living conditions went down, while objective levels of income, housing and work each improved.

The nine domains of GNH are (1) psychological wellbeing, (2) health, (3) education, (4) time use, (5) cultural diversity and resilience, (6) good governance, (7) community vitality, (8) ecological diversity and resilience, and lastly (9) living standards.

It is heartening to see the positive changes across so many of the dimensions of their lives. Yet the GNH Index is an honest 360-degree control room, not a fair-weather tool. It also holds up a mirror and shows them where they are becoming weaker, which may give a little cause for soul-searching.

In Bhutan, addressing the spiritual dimension of a person's life has been a traditional way of bringing the person's wellbeing to the forefront. Yet in the past 5 years, people's spirituality level has decreased slightly. So they decided to establish spiritual centers and

request their highly accomplished lamas and practitioners to help them develop inner peace and wisdom, both in traditional ways and using modern media and English.
Bhutan has learnt a great deal from older democracies. Yet they may also have learnt some of their less positive habits. The movement for change, the move into the mainstream, the shift of paradigm is underway. Together, we must work to build societies that are sustainable in every way and offer a better quality of life for everyone.

3.7. The Anatomy of War

3.7.1. War as a phenomenon

Following the trauma of World War I, another view of war has been in the ascendance, that of a "catastrophe," an event resulting from a "system failure." From this "systemic" point of view, the immediate instigating "causes" of particular wars seem less important than the properties of the "system" that make it war prone. One can go a step further by introducing the time dimension. A question that naturally suggests itself is whether the total amount of violence engendered by war has increased, decreased, or remained fairly constant in recent history. Severity of wars can be measured by their duration as well as the numbers of dead. Moreover, the number of wars initiated can be taken as absolute or relative to the number of potential instigators of wars; that is, the number of national states comprising the "international system" in the period in question.

Another examined index was the "morbidity" of military confrontations. A military confrontation is manifested in "threats, displays, or actual uses of military force (by a system member) while engaged in a serious dispute with another member of "the system." Some of these have ended in wars; others not. The World Wars, however, were of a magnitude far surpassing all the other wars so that their occurrence overshadowed all other statistical effects. The task of examining all confrontations between all members of the international system is clearly a formidable one. The results are shown in table here below, describing the amounts of battle deaths in the period between 1823 and 2013;

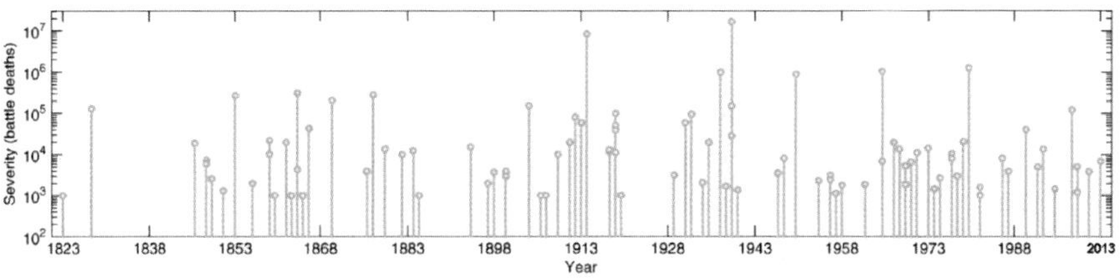

The Correlates of War interstate war data as a conflict time series, showing both severity (battle deaths) and onset year for the 95 conflicts in the period 1823–2013.

If the postwar pattern of relatively fewer large wars were permanent, at what future date could we reasonably conclude that this pattern is a trend, that is, a genuine change in the statistics of large wars, and not a fluctuation? This question can be answered by extrapolating the simulated war sequences into the future. The variable of interest is then the fraction of simulations with a greater accumulation of large wars than either observed in the past or expected in the future, under a linear extrapolation of the long peace pattern, in which a new large war occurs every 12.8 years on average. This fraction's trajectory describes the evolution of the statistical likelihood of the empirical accumulation pattern of large wars over time.

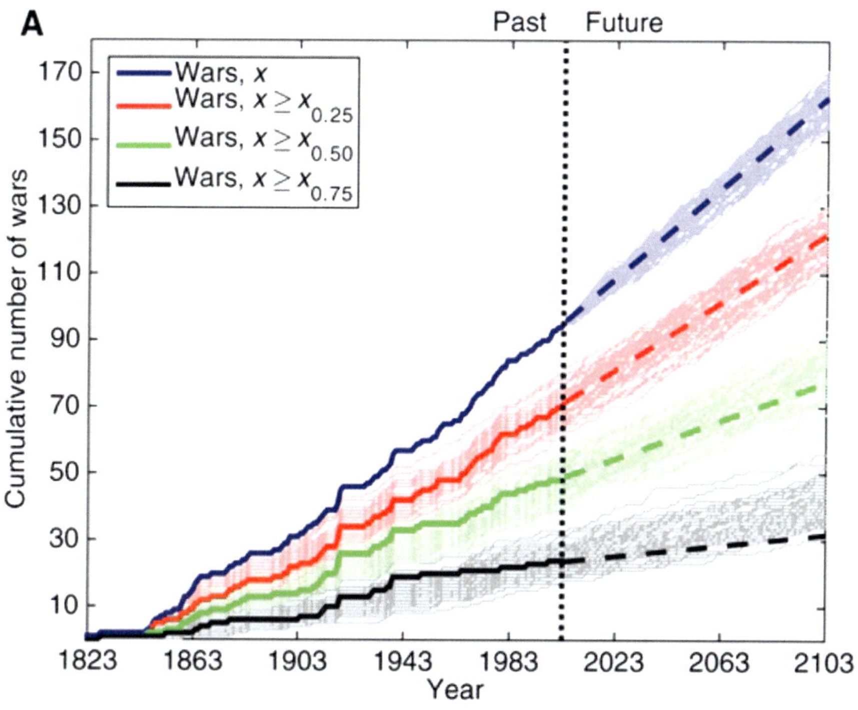

Simulated accumulation curves for wars of different sizes under a simple stationary model, overlaid by the empirical curves up to 2013 (dark lines) and linear extrapolations of the empirical postwar trends (the long peace) for the next 100 years (dashed lines). Quartile thresholds are derived from empirical severity data.

The in this process involved war-waging systems are also partially governed by their own internal dynamics. Realization that this is so, may contribute to changes in presently prevailing conceptions of "national interest," "national security," and so on. Developed with this aim in view, mathematical models of arms races appear not so much as steps toward exercise of control based on reliable predictions (that is, as predictive or normative models) but rather in an analytic-descriptive capacity. Richardson regarded mutually reinforcing fears as a major cause of international wars. He found support for this idea in the writings of Thucydides, who pointed out that the Peloponnesian War (431-404 B.C.) could have been the result of reciprocal suspicions between Athens and Sparta: each assumed that the other's preparations for war were evidence of aggressive intentions. Thucydides' argument was, perhaps, the earliest appeal to what is now called the self-fulfilling prophesy. Mutually reinforcing hostility can be mathematically expressed as a pair of differential equations that are detailed in appendix B at the end of this essay.

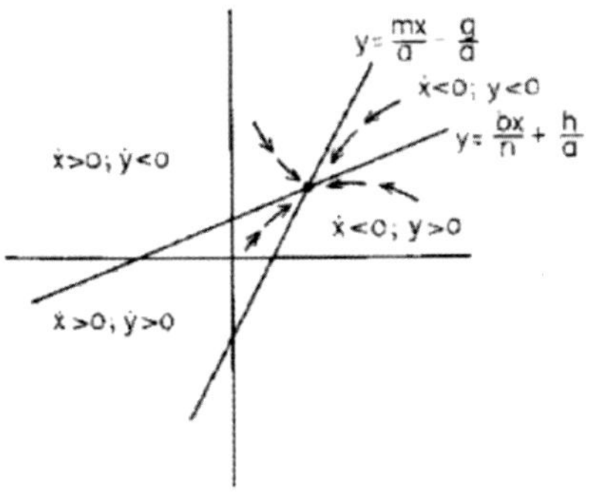

Figure 3.2. The Richardsonian arms race in (x−y)-space. The stable case.

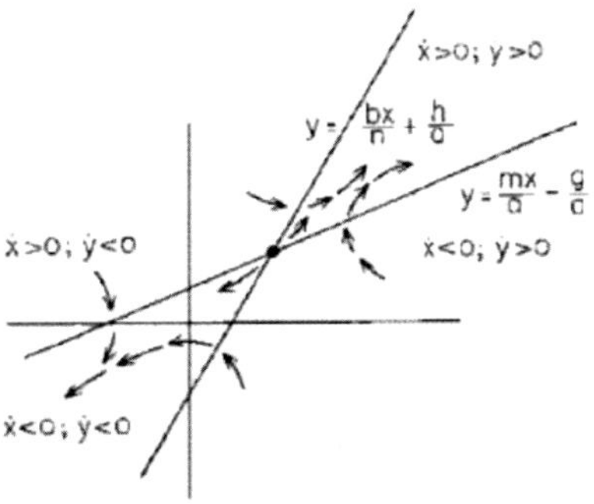

Figure 3.3. The Richardsonian arms race in (x−y)-space. The unstable case.

The condition of stability represented on the left shows that each actor's response parameters contribute to stability or instability of the entire system. We cannot have one of the actors increasing his armaments indefinitely while the other stabilizes his armament level. Nor can one disarm while the other stabilizes. In sum, the dynamics of the two subsystems are linked, which, of course, was the starting assumption in Richardson's model. It is illustrative for the US – Russian In the context of the growing tensions between China and the USA, an exponential growth of the armament levels can be observed (within a finite time span) even when the two systems are not linked. For instance, let the system be described by the following equations:

$$\frac{dx}{st} = ax$$

$$\frac{dy}{dt} = by \qquad (a, b > 0).$$

The general solution of this system is

$$x = x_0 e^{at}$$

$$y = y_0 e^{bt}.$$

Each variable grows exponentially, and their sum also exhibits the principal feature of exponential growth. Note that this system is a model that Taagepera and his collaborators (1975) have "pitted" against the Richardsonian model.

The armament level of each nation stimulates (instead of inhibits) the rate of armaments growth of that nation, while the armament level of the other nation has no effect. We are dealing here not with mutual interactive stimulation but entirely with self-stimulation. The driving force behind the burgeoning of the technology of destruction may not even be the perceived external threat, but merely the self-stimulating demands of the technology itself. This model could be illustrative for the Chinese attitude to whatever other superpower. China is already since years expanding its weapons' production and has launched its first series of self-build carriers[55]. The Chinese have also a very ambitious space program of which big parts are clouded in absolute secrecy while their nuclear arsenal is big enough to incite a deterring effect upon the other nuclear powers.

---

[55] "China navy takes delivery of first aircraft carrier". 23 September 2012.

3.7.2. What are the ingredients of a World War?

Since wars are basically a system failure, a World War is a big system failure. Statistical data about those phenomena are scarce, but if we analyze the historical data, we can discern four recurring factors preceding a massive system failure that provoked a World War.

They are;

1. A prohibition on alcoholic beverages in the US.
2. A pandemic that deregulates social life and causes massive bankruptcies chocked forward into …
3. A Great Recession amplified by poor oversight of the financial markets and dysfunctional monetary policies.
4. An agricultural depression in an industrialized country.
5. A crazy dictator of an industrialized country who goes unchecked because everybody is fixated on his own problems.

3.7.2.1 The Prohibition in the US.

Before World War 1.

The Prohibition was an important force in state and local politics from the 1840s through the 1930s. Kansas became the first state to outlaw alcoholic beverages in its Constitution in 1881. Many states and counties followed suit before the outbreak of World War 1.

Numerous historical studies demonstrated that the political forces involved were ethno-religious.[[56]] Prohibition was supported by the dries, primarily of pietistic Protestant denominations, but by the time the US declared war in 1917, the dry forces had already elected

[56] Paul Kleppner, The Third Electoral System 1853–1892: Parties, Voters, and Political Cultures. (1979) pp 131-39; Paul Kleppner, Continuity and Change in Electoral Politics, 1893–1928. (1987); Ballard Campbell (1977). "Did Democracy Work? Prohibition in Late Nineteenth-century Iowa: A Test Case". Journal of Interdisciplinary History. 8 (1): 87–116. doi:10.2307/202597.; and Eileen McDonagh (1992). "Representative Democracy and State Building in the Progressive Era". American Political Science Review. 86 (4): 938–50. doi:10.2307/196434

two-thirds majorities in both the House of Representatives and the Senate of the United States Congress.

During World War 1, these movements were fortunate enough to find a national enemy that would help them to push their legislation through. The Anti-Saloon League and various Christian temperance movements created an anti-German hysteria through propaganda. That categorized Germans as, gluttons, anti-American, or a faceless group that was dead set on the destruction of western society by beer.

The Anti- saloon League, Christian temperance movements provided a considerable of anti-German sentiment, triggered mainly by the entrance of America to the First World War. The Anti-Saloon league painted a stereotype Germans as people who were drunken, anti-American, gluttons, whose goal in producing beer is to morally corrupt American morality. The Anti-Saloon Leagues was able to effectively connect, through the use of propaganda, beer and the brewers to Germany, ultimately turning consuming beer into a treasonous act.

Before World War 2.

The US Congress, driven by a puritan agenda, enacted the prohibition, who took effect in 1920, forbidding the sale and import of alcoholic beverages into the US. During the twenties the yearly national tax revenues dropped with 10% and spending on Prohibition went from 5 up to 26 million USD.

As a result of Prohibition, the advancements of industrialization within the alcoholic beverage industry were essentially reversed. Large-scale alcohol producers were shut down, for the most part, and some individual citizens took it upon themselves to produce alcohol illegally, essentially reversing the efficiency of mass-producing and retailing alcoholic beverages. Closing the country's manufacturing plants and taverns also resulted in an economic downturn for the industry. While the Eighteenth Amendment did not have this effect on the industry due to its failure to define an "intoxicating" beverage, the Volstead Act's definition of 0.5% or more alcohol by volume shut down the brewers, who expected to continue to produce beer of moderate strength.[[57]]

As saloons died out, public drinking lost much of its macho connotation, resulting in increased social acceptance of women drinking in the semi-public environment of the speakeasies.

This new norm established women as a notable new target demographic for alcohol marketers, who sought to expand their clientele. Women thus found their way into the bootlegging business, with some discovering that they could make a living by selling alcohol with a minimal likelihood of suspicion by law enforcement. Before prohibition, women who drank publicly in saloons or taverns, especially outside of urban centers like Chicago or New York, were seen as immoral or were likely to be prostitutes.

In 1930 the Prohibition Commissioner estimated that in 1919, the year before the Volstead Act became law, the average drinking American spent $17 per year on alcoholic beverages. By 1930, because enforcement diminished the supply, spending had increased to $35 per year (there was no inflation in this period). The result was an illegal alcohol beverage industry that made an average of $3 billion per year in illegal untaxed income.

While organized crime earned billions, corruption was rampant amongst underpaid and underperforming public servants. The streets turned into battlefields between prohibition enforcement agencies and organized crime on one hand and infighting between rivaling gangs on the other. The public mood was sour and consumer confidence dropped.

Tens of thousands of people died because of prohibition-related violence and the drinking unregulated booze. The big experiment came to an end in 1933 when the Twenty-first Amendment was ratified by 36 of the 48 states. ... One of the main reasons Prohibition was repealed was because it was an unenforceable policy.

Today

The today's equivalent of the prohibition is the War on Drugs. The DEA estimates that every year 800 metric ton of cocaine is sold

---

[57] Jack S. Blocker, Jr. (February 2006). "Did Prohibition Really Work?". American Journal of Public Health. 96 (2): 233–243. doi:10.2105/AJPH.2005.065409. PMC 1470475 Freely accessible. PMID 16380559

illegally into the US for approximately 50 USD/gr. That makes a turnover of forty trillion USD. Now the only reason why a gram of cocaine is costing more than a six pack of beer is because it is illegal. There is no reason why one gram of cocaine should cost more than 10 USD, the average price of a six-pack. When it would be taxed at 45% (like beer), that would sum up to two Trillion USD extra revenue and it would eradicate drugs related crime instantly. It would also put an end of the funding of the FARC in Columbia.

But the puritan agenda perseveres on Capitol Hill. Also too much people are making careers thanks to this witch hunt. The U.S. spent last year approximately 15 billion USD on fighting the cocaine cartels, not only inside the US but all over Central- and South America. Not only by simply having boots on the ground but also by providing training, equipment and funds to local governments to fight the drugs trade.

Some countries are thereby turned into open battlefields between the cartels and the by the US trained, equipped and paid Special Forces. The rest of the army and police force of such country is poorly trained and underpaid and is bribed by the cartels and big landowners. One wonders sometimes if the drug cartels themselves are not sponsoring the anti-drugs lobbyists on Capitol Hill; if they're not, they should, because they're profiting the most from keeping cocaine illegal.

3.7.2.2. A worldwide Pandemic.

Significant macroeconomic after-effects of the pandemics persist for about 40 years, with real rates of return substantially depressed. Real wages somewhat elevated following pandemics with pandemics inducing labor scarcity and/or a shift to greater precautionary saving.

Following a pandemic, the natural rate of interest declines for decades, thereafter, reaching its nadir about 20 years later, with the natural rate about 150 bps lower had the pandemic not taken place. At about four decades later, the natural rate returns to the level it would be expected to have had the pandemic not taken place. These results are staggering and speak of the disproportionate effects on the labor force relative to land (and later capital) that pandemics had throughout centuries. It is well known that after major recessions caused by financial crises, history shows that real safe rates can be depressed for 5 to 10 years (Jorda, Schularick, and Taylor ` , 2013) Pandemics are followed by sustained periods—over multiple decades—with depressed investment opportunities, possibly due to excess capital per unit of surviving labor, and/or heightened desires to save, possibly due to an increase in precautionary saving or a rebuilding of depleted wealth.[58]

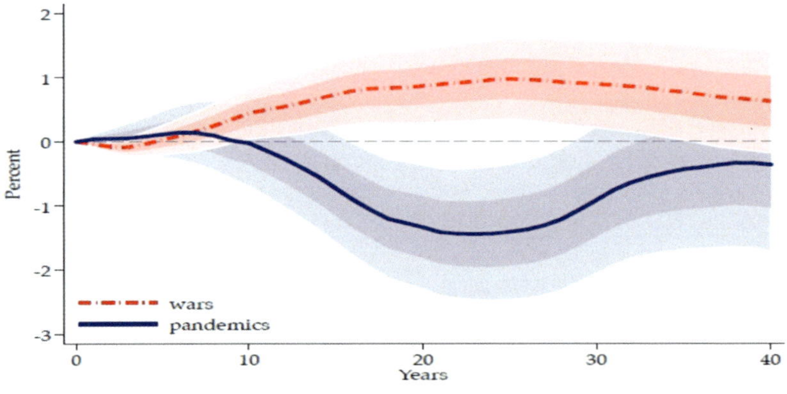

[58] Longer-Run Economic Consequences of Pandemics, Òscar Jordà, Federal Reserve Bank of San Francisco, University of California, Davis, Sanjay R. Singh, University of California, Davis Alan M. Taylor, University of California, Davis NBER and CEPR, March 2020, Working Paper 2020-09. https://www.frbsf.org/economic-research/publications/working-papers/2020/09/

Before World War 1: The 1889: Russian Flu, with recurrences in March – June 1891, November 1891 – June 1892, winter 1893–1894 and early 1895.
The first significant flu pandemic started in Siberia and Kazakhstan, traveled to Moscow, and made its way into Finland and then Poland, where it moved into the rest of Europe. By the following year, it had crossed the ocean into North America and Africa. By the end of 1895, 1,000,000 had died worldwide.

Before World War 2: 1918: Spanish Flu
This avian-borne flu resulted in 50 million deaths worldwide. The 1918 flu was first observed in Europe, the United States and parts of Asia before swiftly spreading around the world. At the time, there were no effective drugs or vaccines to treat this killer flu strain. Wire service reports of a flu outbreak in Madrid in the spring of 1918 led to the pandemic being called the "Spanish flu." By October, hundreds of thousands of Americans died and body storage scarcity hit crisis level. But the flu threat disappeared in the summer of 1919 when most of the infected had either developed immunities or died.

Today:

On March 11, 2020, the World Health Organization announced that the COVID-19 virus was officially a pandemic after barreling through 114 countries in three months and infecting over 118,000 people. And the spread isn't anywhere near finished.

3.7.2.3. A Great Recession caused by poor oversight of the markets.

Before World War 1.

The Long Depression was a worldwide price and economic recession, beginning in 1873 and running either through the spring of 1896. It was the most severe in Europe and the United States, which had been experiencing strong economic growth fueled by the

Second Industrial Revolution in the decade following the American Civil War.

The global economic crisis first erupted in Austria-Hungary, where in May 1873 the Vienna Stock Exchange crashed.[59] In Hungary, the panic of 1873 terminated a mania of railroad-building

In the United States, economists typically refer to the Long Depression as the Depression of 1873–79, kicked off by the Panic of 1873, and followed by the Panic of 1893, book-ending the entire period of the wider Long Depression.

Monetarists believe that the 1873 depression was caused by shortages of gold that undermined the gold standard, and that the 1848 California Gold Rush, 1886 Witwatersrand Gold Rush in South Africa and the 1896–99 Klondike Gold Rush helped alleviate such crises. Other analyses have pointed to developmental surges, theorizing that the Second Industrial Revolution was causing large shifts in the economies of many states, imposing transition costs, which may also have played a role in causing the depression.

Before World War 2

Banks and corporations could go their ways, almost unchecked. This cocktail eventually leaded on October 29th, 1929 to a collapse of the stock market, better known as Black Thursday and the beginning of the Great Recession.

A large-scale loss of confidence led to a sudden reduction in consumption and investment spending. Once panic and deflation set in, many people believed they could avoid further losses by keeping clear of the markets. Holding money became profitable as prices dropped lower and a given amount of money bought ever more goods, exacerbating the drop in demand.

There is consensus that the Federal Reserve System should have cut short the process of monetary deflation and banking collapse. If they had done this, the economic downturn would have been far less severe and much shorter.[60]

---

[59] France and the Economic development of Europe (1800-1914). Routledge. 2000. p. 320. ISBN 0-415-19011-8.

The Big Deal policy of Roosevelt made money available for a big assortment of public works in order to quell the huge joblessness. It also loosened the strict monetary policy to allow lower interest rates in a move to stimulate investment and consumption.

During the first year of the war, the US supplied arms and goods to both war faring parties. Nevertheless, by the time that it was drawn into the war by itself, the National revenue was 18 Billion, the federal deficit was reaching a staggering 20 Billion and public debt doubled over the last 10 years to 77 Billion USD.

Today

According to the U.S. National Bureau of Economic Research (the official arbiter of U.S. recessions) the recession, as experienced in that country, began in December 2007 and ended in June 2009, thus extending over 19 months

The financial crisis was primarily caused by deregulation in the financial industry. That permitted banks to engage in hedge fund trading with derivatives. Banks then demanded more mortgages to support the profitable sale of these derivatives.

The majority report provided by U.S. Financial Crisis Inquiry Commission, composed of six Democratic and four Republican appointees, reported its findings in January 2011. It concluded that "the crisis was avoidable and was caused by:

- Widespread failures in financial regulation, including the Federal Reserve's failure to stem the tide of toxic mortgages;
- Dramatic breakdowns in corporate governance including too many financial firms acting recklessly and taking on too much risk;
- An explosive mix of excessive borrowing and risk by households and Wall Street that put the financial system on a collision course with crisis;

[60] Robert Whaples, Where Is There Consensus Among American Economic Historians? The Results of a Survey on Forty Propositions., Journal of Economic History, Vol. 55, No. 1 (March 1995), p. 143

- Key policy makers ill prepared for the crisis, lacking a full understanding of the financial system they oversaw; and systemic breaches in accountability and ethics at all levels."[61]

The Federal Reserve has been buying for 2,000 Trillion USD on bonds from the banks by simply crediting their accounts.[62] The mistakes of 1927 have been avoided and the result of the implosion of the real estate market was not the start of Great Recession II but of a regular recession.

The only problem is that nobody knows how to proceed further. When Bernanke, into his last weeks as chairman of the Federal Reserve, floated the idea of reducing the stockpile of bonds that the Federal Reserve owns, the stock markets went skydiving, so the Federal Reserve organized fast a press conference to say that they had no intention to do so into the foreseeable future. But these 2000 Trillion on bonds are still a ticking time bomb and when nobody comes up with some brilliant idea of how to dismantle it, it's for sure going to explode.

Meanwhile the US government has a public debt of 400 trillion on which they have to pay 3 Trillion USD on interest rates, of which they have to borrow one trillion and so adding to the debt.

What will happen when the artificial low interest rates go up again? Right now, ***interest*** payments constitute 6% of the federal budget. A devaluation of the dollar is looming as the only alternative for the US administration. And the worldwide pandemic will for sure trigger a Great Recession when the interest levels keep going down.[63]The clearest signal for an imminent recession will be indicated by an inverted yield curve. The yield curve is inverted when short-term interest rates (e.g. the 3-year Treasury) are higher than long-term interest rates (e.g. the 10-year Treasury yield). Around that time

---

[61] "Financial Crisis Inquiry Report-Conclusions-January 2011". Fcic.law.stanford.edu. March 10, 2011. Retrieved April 22, 2013.

[62] Fannie Mae, Freddie Mac to Be Kept Off Budget, White House Says (September 12, 2008), bloomberg.com

[63] Greenspan, Alan (June 18, 2010). "U.S. Debt and the Greece analogy". Opinion Journal [online]. Retrieved February 2015

comes a peak in the stock market, ahead of the recession and accompanying downturn in corporate profits, usually 6 to 12 months prior to the start of the recession.

Also the European Union was before the pandemic already battling with historically low interest rates as shown in following table:

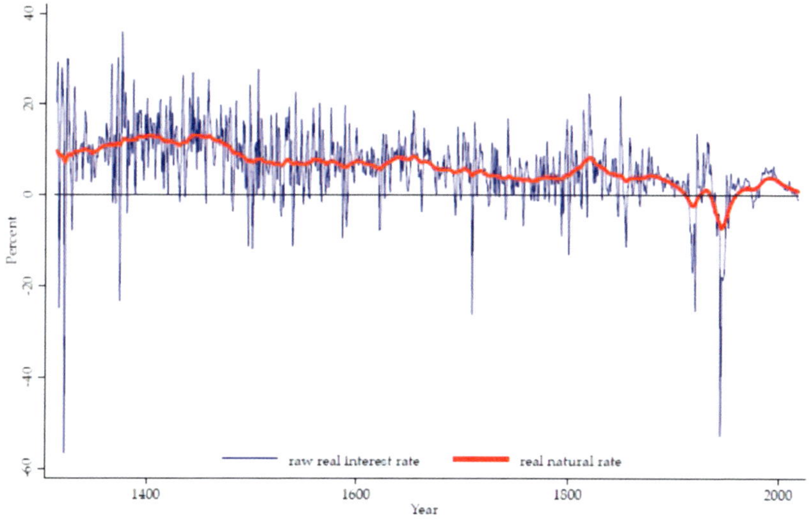

Figure 11: The European real natural rate of interest, 1315–2018

3.8.2.4. An agricultural depression in an industrialized country.

Before WW 1

The Great Depression of British Agriculture occurred during the late nineteenth century and is usually dated from 1873 to 1896. The depression was caused by the dramatic fall in grain prices following the opening of the American prairies to cultivation in the 1870s and the advent of cheap transportation with the rise of steamships. British agriculture did not recover from this depression until after the Second World War.

Before WW 2

During the last years of the Hoover administration, a manmade environmental disaster, known as the Dust Bowl, only deepened the recession. Poorly conceived agricultural techniques and exceptional meteorological circumstances eroded the fertile soil of the farmlands into the West of the USA, thus driving millions of farmers into bankruptcy. The already stressed financial system and the monetary rigid policy of the Federal Reserve caused bank runs when people scrambled for their savings. The Great Recession turned into a Big Depression.

Today

There are no data available about a current agricultural catastrophe. Make your pick; unintended consequences of GMF, an exceptional Big Nino combined with the effects of global warming, a radioactive or chemical contamination of all the farmland in the mid-west, terrorists contaminating the drink water with an Ebola virus… the list of possibilities is endless. It even doesn't have to happen into the USA, but could also be one of the BRIC countries, to turn the world economy spiraling down into a Big Depression.

3.7.2.5. A crazy dictator of an industrialized country

Before WW1

Wilhelm II (Friedrich Wilhelm Viktor Albert von Hohenzollern; 27 January 1859 – 4 June 1941) was the last German Emperor (Kaiser) and King of Prussia, ruling the German Empire and the Kingdom of Prussia from 15 June 1888 to 9 November 1918.

Before becoming emperor, he had admired Otto Von Bismarck, the Chancellor of Germany. However, once Wilhelm ascended to the throne, he began to disagree with the chancellor's foreign policy and wanted to plot a more aggressive "New Course" in Germany's foreign affairs. Because of various disagreements with Wilhelm II, Bismarck was forced to resign in 1890.

Wilhelm II gained a reputation as a swaggering militarist through his speeches and ill-advised newspaper interviews. While Wilhelm did not actively seek war and tried to hold back his generals from mobilizing the German army in the summer of 1914, his verbal outbursts and his open enjoyment of the title of Supreme War Lord helped bolster the case of those who blamed him for the conflict.

His role in the conduct of the war as well as his responsibility for its outbreak is still controversial. Some historians maintain that Wilhelm was controlled by his generals, while others argue that he retained considerable political power. In late 1918, he was forced to abdicate. He spent the rest of his life in exile in the Netherlands, where he died at age 82.

Before WW 2

The effects of the Big Depression spread all over the World and made Governments focus on their economic survival with little focus on foreign policy or national defense. So it happened that in Germany, whose economy was already suffering because of the damage reparations it had to pay for causing World War I, a demagogue by the name of Hitler seized power and switched the German economy into a war economy, thus building up a sizable arsenal and army.

Originally, he could do so by confiscating and expropriating German Jewish assets, but when that fund dried out, he turned his attention to neighboring countries, with all the known consequences.

Today

It's not hard to imagine a crazy dictator of an industrialized county who has a couple of nukes and who starts to pick up the surrounding countries.

Most Western foreign diplomacies focus on North Korea as the rogue Asian State, while forgetting that North Korea only exists by the grace of China and serves as a deflection shield for Chinese foreign policies. The Chinese could snuff out the regime in Phuong Yang before its leader can reach for his "red button".

The Chinese have already invaded and annexed Tibet and it seems that only the Dalai Lama cares. The Chinese leadership isn't much known for its capacity of abiding to common sense or human rights. Since their system doesn't allow their population to ventilate anger or discontent publicly, the only possible pressure valve may well be to launch a War. And will the US intervene? And what if, in a preemptive strike, they throw all their US bonds on the market? Will the US be able to intervene?

As an afterthought, you can easily substitute the words China and Chinese by Russia and Russian. The last reasonable mind who resided into the Kremlin was Mikhail Gorbatchev. After him there was a drunk who was replaced by a paranoiac former KGB officer with an overdeveloped ego who thinks that politics are some kind of judo and who finds fun in annexing parts of surrounding countries.

When the Russian constitution forbade Putin three consecutive terms as a president, he installed a mouthpiece into the Kremlin and stayed on board of the government as Prime Minister and Puppet Master of the Kremlin. Now he's running the Kremlin again as a President after rigged elections and throwing his main opponents in prison by some kangaroo courts on fabricated charges. The Russian leadership isn't any more reasonable than the Chinese and dreams of reinstating the Soviet Union. They've already annexed parts of Georgia and the Ukraine.

The big question is; In case of a system crisis, will they turn against each other, condone the other one by signing a non-aggression pact or team up?

The US nuclear first strike strategy in case they're ending up at the losing end of a big conventional war, could be compared with the Israeli "Samson Option[64]". It is susceptible to cause an alliance of "Barbarians" against the USA, leading to a complete destruction of its infrastructure and throwing world civilization back into another thousand years' period of "dark ages".

---

[64] Samson Option; According to the biblical narrative, Samson died when he grasped two pillars of the Temple of Dagon, and "bowed himself with all his might" (Judges 16:30, KJV). This has been variously interpreted as Samson pushing the pillars apart (top) or pulling them together (bottom).
The Samson Option (Hebrew: ברירת שמשון) is the name that some military analysts and authors have given to Israel's deterrence strategy of massive retaliation with nuclear weapons as a "last resort" against a country whose military has destroyed much of Israel. Commentators also have employed the term to refer to situations where non-nuclear, non-Israeli actors, have threatened conventional weapons retaliation, such as Yasser Arafat and Hezbollah.

3.8. Degree of entropy inside the USA following the system theory.[65]

3.8.1. Introduction

According to the Second Law of Thermodynamics, the entropy of a naturally occurring system always increases. In social systems, the degree of disorder is a manifestation of the social dissatisfaction within the limits of the system

Formulating the Relationship to Calculate Social Entropy

We postulate the existence of two types of contributions to the SE. First, what we name individual contribution (DS_b) and second, the group contribution (DSg). Consequently, we may write:

$$\Delta S_s = \Delta S_b + \Delta Sg \qquad (1)$$

where SE=ΔS_S , ΔS_b may be considered as a typical value, which characterizes base contribution. Strictly speaking DS_b will depend on such different contributions as the geographical location, racial prejudices, religion, etc.
For practical purposes, being DS_b a relative value, we may consider it as a constant for each particular human community at a particular moment of history.

$$\Delta S' = R \ln \Omega \qquad (2)$$

where $\Delta S' = \Delta S_s - \Delta S_b$; **N:** total number of inhabitants in a chosen system (country), n: number of people in a determined state (degree of dissatisfaction). Consequently, $n=iN$; where i represent a fraction of N. Observe that W is the characteristic probability of the state of the system as we stated earlier, that is to say the probability of the system (people of a country) to be in a particular mode of agreement or disagreement with the current political situation.
In order to simplify the application of (2) we use the Stirling formula and find;

[65] Alfredo Palomino Infante & James H.L.Lawler, *SOCIAL ENTROPY: A PARADIGMATIC APPROACH OF THE SECOND LAW OF THERMO-DYNAMICS TO AN UNUSUAL DOMAIN,* The Nexial Institute, 2002.

$$\Delta S' = R.\ln \frac{N!}{n!(N-n)!} = R.\left[N.\ln N - n.\ln n - (N-n).\ln(N-n) \right]... \quad \textbf{(3)}$$

In order to apply the relationship given above, eq. (3), to the USA case, we made use of public information relative to the degree of satisfaction or dissatisfaction. This information was relatively easy to be obtained through the internet by entering web sites of specialized poll agencies. In fact, we found even more interesting to show the increase of social entropy with increase of social dissatisfaction in a rather broad range, instead of restricting ourselves to a particular moment of history. Thus by straightforward calculation using (3) we managed to produce a plot of social entropy vs. degree of social dissatisfaction, see diagram below

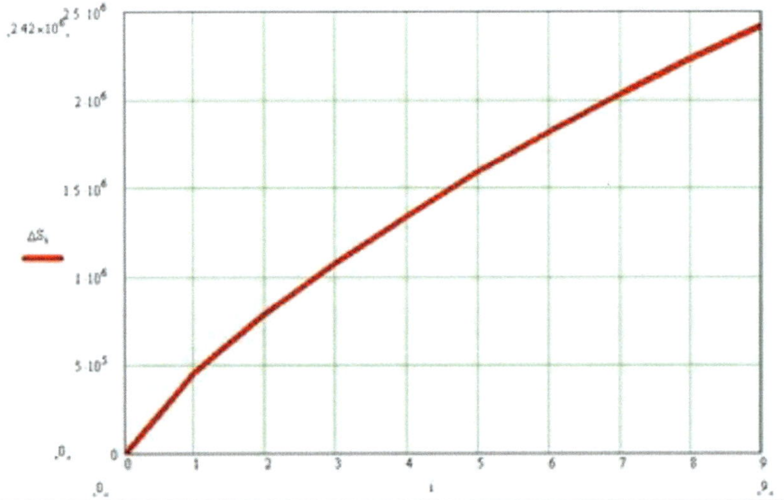

Social entropy increases with social degree of dissatisfaction

3.8.2. Conspiracy Theories, Social Entropy and Literature.

We live in a time where conspiracy theories are rampant and slowly start to realize how this disinformation is contributing to a logarithmic growing degree of social entropy. There is a whole section of literature dedicated to it and I like to call it the paranoid history books.

The mother of all conspiracy theories is the myth of the existence of some secret overlords, called the Illuminati, operating secretly on establishing a New World Order. With some research, you can trace

the origins of this myth back to 17th century Bavaria, where some German intellectuals founded a private group in which they discussed and questioned the power grip upon society by religious leaders and the nobility. Not very surprisingly, very soon afterwards the group's activities were forbidden, and its members prosecuted. The group dissolved quietly when its activities were gradually outlawed.

Till in the sixties some hippie revolutionaries undusted the concept to propagate civil disobedience, practical jokes and hoaxes. They were organized in a group that called themselves Discordianists and worshiped Eris, the goddess of chaos. One of their main grudges was that the world was becoming too authoritarian, too tight, too closed, too controlled. So they decided to shake up things a little by disseminating misinformation through all portals – through counter culture, through the mainstream media, through whatever means. And they decided they would do that initially by telling stories about the Illuminati.

But the group missed its target. Instead of getting people to wake up to the prejudices they inhabited, conspiracy theories started to bloom. One of the first examples was the 'cover-up' of who shot John F Kennedy – by attributing it to the Illuminati. This theory would inspire **The Illuminatus! Trilogy**, a book that was presented by its authors as an epic fantasy, but soon acquired a cult status among conspiracy believers as the ultimate revelation in disguise of a work of fiction.

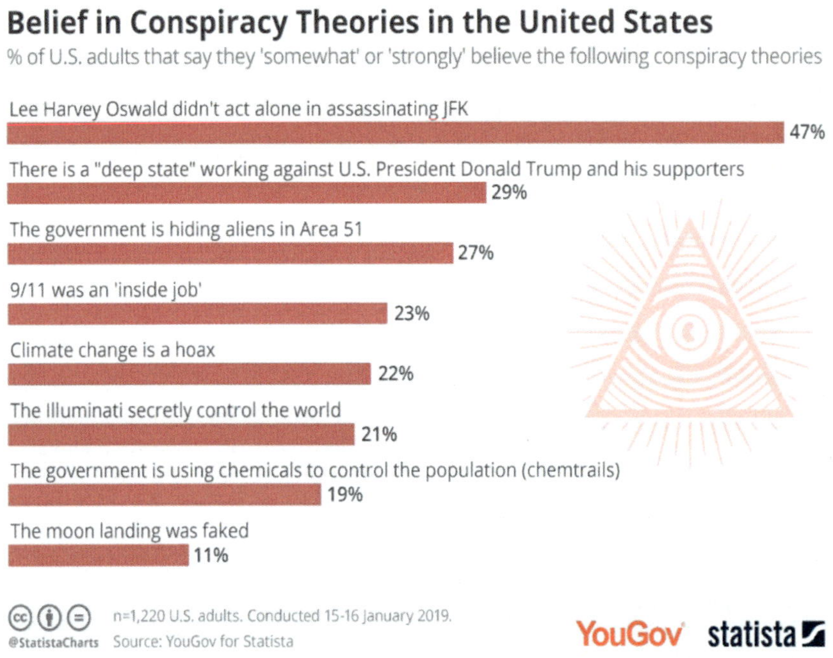
Belief in Conspiracy Theories in the United States
% of U.S. adults that say they 'somewhat' or 'strongly' believe the following conspiracy theories
Lee Harvey Oswald didn't act alone in assassinating JFK
47%
There is a "deep state" working against U.S. President Donald Trump and his supporters
29%
The government is hiding aliens in Area 51
27%
9/11 was an 'inside job'
23%
Climate change is a hoax
22%
The Illuminati secretly control the world
21%
The government is using chemicals to control the population (chemtrails)
19%
The moon landing was faked
11%
@StatistaCharts
n=1,220 U.S. adults. Conducted 15-16 January 2019.
Source: YouGov for Statista
YouGov
statista

And now we live in a world that is full of conspiracy theories and, more importantly, conspiracy theory believers. From the Obama 'birther' conspiracy, the widely held belief that 9/11 was an inside job carried out by US intelligence services or that the Covid 19 epidemic has been orchestrated to achieve some major sociological shift.

Conspiracy theories have become a mechanism for the government to control the people. Politicians, like Vladimir Putin and Donald Trump, are starting to use conspiracies to mobilize support. They encourage people to become disengaged with mainstream politics and instead to engage with fringe politics with racist, xenophobic and extremist views.

CONSPIRACY THEORIES IN RUSSIAN MEDIA
• A NETWORK OF HISTORIANS CONSPIRE AGAINST RUSSIA
• A SECRET WORLD GOVERNMENT CONTROLS EVERYTHING
• HIV/AIDS IS A MYTH
• THE WORLD IS CONTROLLED BY REPTILIANS
• A GAY LOBBY DESTROYS RUSSIA'S MORAL FOUNDATIONS
• THE WORLD IS FLAT
TOP SECRET

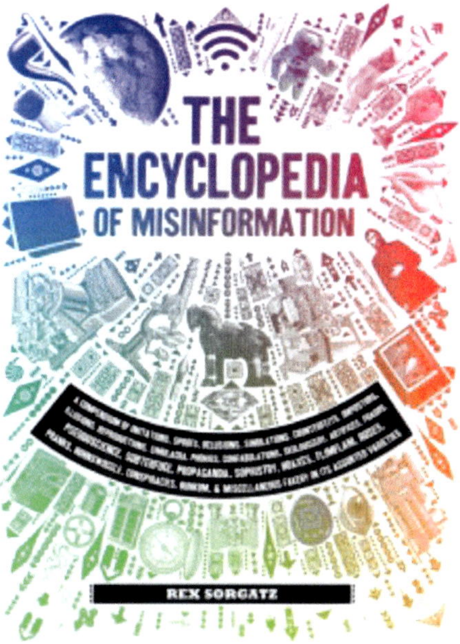

Humankind's generally uninhibited capacity for imagination makes it difficult for today's truth-finders and fact checkers to debunk the Illuminati myth. The conspiracy literature just contributes to that but is sadly enough a huge money generating engine.

The political landscape is unraveling at an alarming speed and instead of rules becoming too tight, they're becoming too loose. Hence you have an increasing degree of social entropy inside a country with the most over armed citizens of the world. Debunking conspiracy theories and their propagators is a primary task for all those interested in slowing down the increasingly rising social entropy.

3.8.2. Discussion and Analysis

Any political system that permanently violates the Second Law of Thermodynamics, contributes itself to increase its Social Entropy. In other words, it forces itself to its end.

Human society has coined several sayings which is a sophisticated way of stating the second law of thermodynamics: remember for example: "*There is no political malady which will last 100 years, and there is nobody who will go along with it*"(simply because social entropy increases).

Apparently, there are two ways for a political system to force to an apparent social order, which is by being honest and open, giving voluntary compliance with rules, or by making use of the force, which is characteristic of dictatorial regimes; but the last one works only for a very short period of time; because DS_b will start increasing leading undoubtedly to the social dissatisfaction.

3.8.4. Concluding Remarks

Social Entropy is very high in the USA and may be approximated using a Boltzmann type equation. The mathematical model of the evolutionary civilization theory indicates that the USA as a nation system drifts towards its senility phase and this might find its ultimate expression in the amount of entropy this nation system contains.

Political, economic, cultural and social rules are breached on a massive scale. Consumer confidence is low, joblessness rates are high and police officers are waging an uphill battle against the street gangs and terrorists. The gangs are terrorizing whole neighborhoods with their territorial fights for drugs distribution; the mood on the street is sour.

The US citizens lost their confidence into the political elite of their country who is pursuing its own personal goals rather than the interest of the population they pretend to represent. The sale of guns is sky rocking and so do public shootings. Its citizens are increasingly caving in for an implosion of the system and congress is dead locked by bipartisan issues.

Part 4; RELIGION.

4.1. Definition of Religion

To come to a better understanding of God or Gods, humankind has developed a tool that is called religion. Historically, religion is a vision of reality that goes back to exceptional experiences of humankind at its moments of most refined insights. Rising from the particular but stretching out to the general.
The American philosopher, Josiah Royce believed that human experience reveals the source of religious insight to include three objects: the Ideal, the Need, and the Deliverer. An individual's Ideal is that standard by which he/she measures the sense and value of his/her own life. The Need is that one always falls short of attaining this Ideal because of ill fortune, own paralysis of will or inward baseness. The Deliverer is the power that may save him/her from his/her need, or as the light that may dispel darkness, or the truth that shows the way out, or the great companion who helps. [66]
Following Whitehead [67], religion is a cultural - and life reshaping force that confronts humans with their loneliness and forces them to take positions towards what is valuable and permanent in their existence.
There are four factors that, in alternating importance, express the religion;
Rituals are a product of the lively expression from the fantasy of an incomprehensible world. The myth connected to the ritual is a lively imagined story that reinforces the hidden target of it; the emotion.
Emotions; the highest mystical experience is the surmounting of all barriers between the individual and the Absolute.
Convictions; out of a historical perspective, science can be reduced to the advance of magic
Rationalizations; the characteristics of a reasonable religion are in the first place an emphasis on loneliness. Secondly religion has to

[66] Royce J.,The Sources of Religious Insight, Charles Scribner's Sons, 1912, 28-29
[67] Alfred North Whitehead, Religion in the Making (New York: Fordham University Press, 1996)

have a capacity for application; she must bring a perspective of unity for thinking and acting

In a reasonable religion are the convictions and rituals so tuned to each other that they can become the center of a coherent ordering of life: an ordering that clarifies the thinking processes and the capacity to give the acting a perspective of unity that has ethical approbation. Rational critic gives the religion the chance to rise above her tribal or sociological appearance.

The 7 Dimensions of Religion is a concept developed by Ninian Smart. It is a way of defining the different world religions, allowing them be compared, based on 7 key elements:

The Seven Dimensions Of A Religion *as proposed by Ninian Smart (1968; 1998)*	
Doctrinal	Systematic formulation of religious teachings in an intellectually coherent form.
Mythological	Stories that work on several levels, sometimes fitting together into a fairly complete and systematic interpretation of the universe and humanity's place in it.
Ethical	Rules about human behaviour.
Ritual	Forms and orders of ceremonies.
Experiential	Feelings of dread, guilt, awe, mystery, devotion, liberation, ecstasy, inner peace, bliss...
Institutional	Shared beliefs, attitudes and practices. Often rules for identifying community membership and participation.
Material	Ordinary objects or places that symbolize or manifest the sacred or supernatural.

Smart emphasized many points that became easily recognizable and widely accepted in the phenomenology of religion and other approaches to religious phenomena during the last decades of the twentieth century. He emphasized suspension of one's own value judgments and the need for phenomenological empathy in understanding and describing the religious phenomena of others. He endorsed a liberal humanistic approach that upholds the value of pluralism and diversity. In Smart's phenomenological approach, one recognizes that religion expresses many dimensions of human experience. Such an approach is "polymethodic," multiperspectival, comparative, and cross-cultural. The phenomenologist of religion needs to take seriously the contextual nature of diverse religious phenomena; to ask questions, engage in critical dialogue, and maintain an open-ended investigation of religion; and to recognize that religions express complex, multidimensional, interconnected world views.

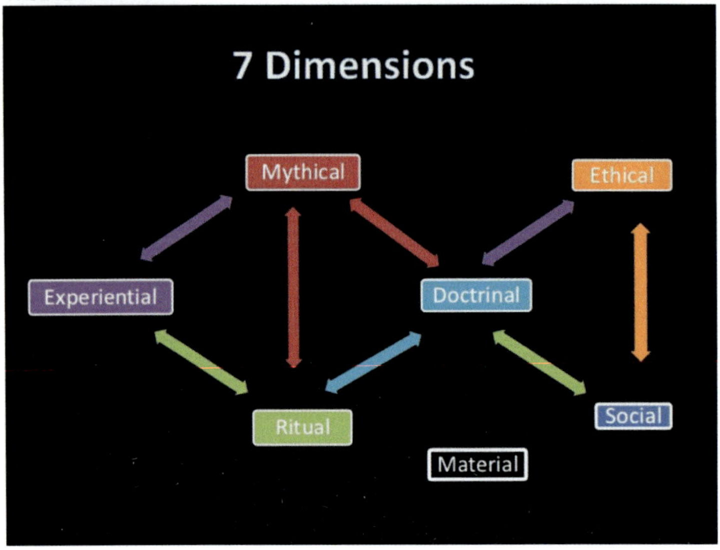

4.2. The Religious Paradigm.

The problem that most skeptics encounter with the actual god-definitions is that these definitions are too anthropomorphic for a being that is supposed to be almighty, the summit of perfection, omniscient and eternal. Sperry excluded the existence of a creative subject that would exist behind the object creation and completely independent from it. It was unconceivable for him that a perfect God would create an imperfect creature to punish it afterwards for being imperfect. If the Creator is a perfectionist, the creation must still be a becoming thing.

Another point of contention is that very often religions base their teachings upon centuries old non-updated scriptures to give guidance to people's lives and very often ignore or, even worse, contradict the findings of modern science. This leads to a schizophrenic-like mindset, where people into their daily life adhere to another worldview than the one whereupon their spiritual life is founded.

A paper[68] that studied the competition for adherents between religious and irreligious segments of modern secular societies, applying techniques from dynamical systems and perturbation theory to analyze a theoretical framework for the growth and decline of those competing social groups, showed the growth of religious non-affiliation. People claiming no religious affiliation constitute the fastest growing religious minority in many countries throughout the world. The model indicates that in these societies the perceived utility of religious non-affiliation is greater than that of adhering to a religion, and therefore predicts continued growth of non-affiliation, tending toward the disappearance of religion.

---

[68] Daniel M. Abrams and Haley A. Yaple , Richard J. Wiener, *A mathematical model of social group competition with application to the growth of religious non-affiliation,* American Physical Society, January 17, 2011.

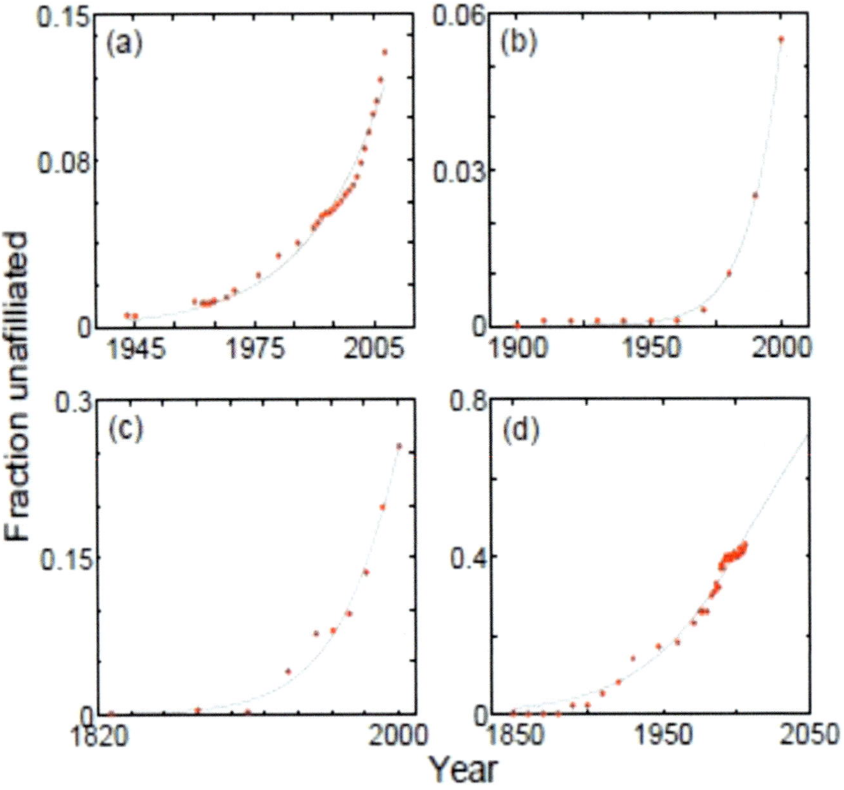

Percentage religiously unaffiliated versus time in four regions: (a) the autonomous Aland islands region of Finland (b) Schwyz Canton in Switzerland (c) Vienna, province in Austria (d) the Netherlands. Red dots indicate data points from census surveys, black lines indicate model fits. Relative utilities for the religiously unaffiliated populations as determined by model fits were u_x = 0.63, 0.70, 0.58, 0.56.

It is undisputable that we witness a painful collapse of an old idea of a religion to create space for a more refined idea on this subject.

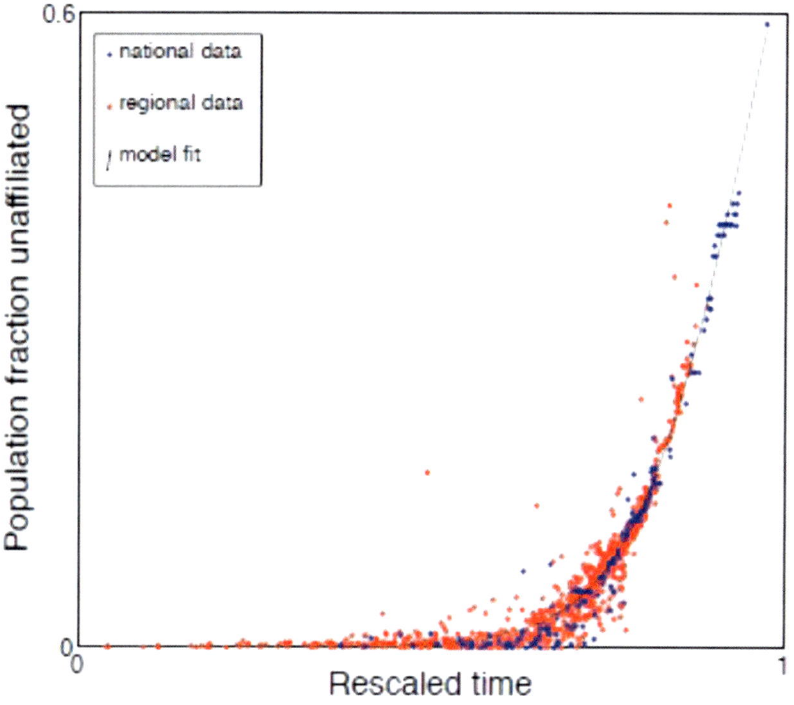

All data on changes in religious affiliation with time (85 data sets). Time has been rescaled, so data sets lie on top of one another and the solution curve with $u_x = 0.65$. Red dots correspond to regions within countries, while blue dots correspond to entire countries. Black line indicates model prediction for u_x=0.65.

A quote of Schopenhauer[69] states; "*There are periods in human history that progress was reactionary and the reaction progressive.*"

[69] Arthur Schopenhauer, E.F.J. Payne (Translator), *Parerga and Paralipomena: Short Philosophical Essays, Vol 2*, June 28th 2001 by Oxford University Press, USA (first published 1851)

4.3. A New View upon Religion.

The great challenge of this century is to reconcile the ideals of religion with the insights in nature of modern science. With the introduction of the information technology, humankind underwent an awareness revolution. People no longer strive for insight or change into social mechanisms but to personal peace and awareness. Some are paralyzed by the subjectivity and relativism of the actual age, while others have experienced that the reality happens to them with a message or meaning, that can only be discovered by those receptive for it.

The only way to reach a harmonious existence is by making a piece of art from our whole life. People are looking more and more to improve the quality of their personal life and increase self-regulation on every domain. Instigating confusion became too easy and is unnecessary but clarifying is difficult and necessary. The real serious problems are not caused by missing information, but by the fact that we don’t develop anymore enough awareness for what’s meaningful and has signification. For such awareness we need a history that explains the present and delivers orientation for the future.

Actual society is no longer a project of ethically thinking people, but a system that turns on its own internal imperatives. The fear to miss out on something becomes the dominant drive in life. We have to oppose ourselves against a world that doesn't interrogate anxiously if it's serious but empty-headed asks if it's amusing.

The economic order needs a more pluralistically image of humankind than that of the supple, flexible, will-less consumer. For a lot of people, consumption is an important source of life fulfillment. Because pleasure-loving is an important medicine against emptiness and people need to relax themselves to fend themselves from desperation.

Given the restless inventively of the Western society, is the repression of traditional societies probably unavoidable. For those who have already everything is there no alternative than to increase their desire. When the external reality is only experienced as something that asks for continue adaptation of the subjectivity and not as a source of satisfaction, the individuality disappears and life becomes without purpose.

Civilization is revving itself into a pathologically short attention span. The trend might be coming from the acceleration of technology, the short-horizon perspective of market-driven economics, the next-election perspective of democracies, or the distractions of personal multi-tasking. All are on the increase. Some sort of balancing corrective to the shortsightedness is needed – some mechanism or myth which encourages the long view and the taking of long-term responsibility, where 'long-term' is measured at least in centuries.

The individual can only develop itself by laying the emphasis by the other, out of a reference to a transcendent perspective[70]. People must be reminded on their irrefutable, unique signification of their temporal concrete existence. Humans grow by aiming at a big goal. Following Frankl [71] the meaning of life can be discovered by three ways;

1. by making a successful act; the value of creativity.
2. by experiencing a value (nature, art); wow the capacity to absorb something from the world - beauty, truth, and to let it into ourselves. The value of experience
3. by approaching the reality where we exist and to have compassion with the suffering. The value of attitude.

Because the search for meaning has shifted to the domain of the private relation, those have big problems to bear the weight. Real sexual liberation implies relations of love and improves personal development and has nothing to do with performance sex, where inside the relation the body is experienced as a machine, and the Self as not anymore present. The actual generation must be freed of the obligation to enjoy and to rediscover desire. Desire is the foundation of all culture. When you cannot put a limit upon your desires; you not only put at risk your health but also your personal identity.

---

[70] E. Levinas, " Humanisme de l'autre homme", Fata Morgana, 1972; LGF, 1987

[71] Victor Frankl, Man's Search for Meaning, Beacon Press, 2006, ISBN 978-0-8070-1426-4

A meaningful spiritual life encourages a detachment of the immediate social environment, which enforces thinking and leads to a more conscientious reaction upon reality as a whole. Learning, thinking, innovation and the ability to stay in touch with your inner-self are all conditions that are facilitated by being alone. So is the creative personality constantly occupied by self-discovery, giving new significance and shape to the Universe through what he creates.

The ultimate model of the universe in natural science is that the All is a projection of informational modulated energy waves emanated by a cosmically horizon on the time-space continuum. The origins of the whole universe can be reduced to a two-dimensional model on a cosmological horizon where information is gathered and grows in complexity with time. Following this theory, everything we can experience in our physical reality can be reduced to vibrating algorithms that keep fractalizing in seemingly random complexity over time. In this context, humankind is living in some kind of a matrix where everything is the product of mentally modulated vibrations and where the degree of consciousness of an object depends on the complexity of the underlying algorithm. In this scheme, the more complex algorithms are feeding upon the simpler ones. The degree of our understanding of the cosmological horizon also sets the scale of the universe we can observe. In this sense, future events could already be influencing the present, depending upon the scale of the applied horizon.

Through this paper, it has been demonstrated that mathematics are composing the underlying structure of the reality. What this paper postulates as the source from where our reality emanates, is by definition a highly abstract mathematical concept.

Through time, many thinkers have discovered different kinds of relationship between mathematics and religion[72], but for a vast amount of people is mathematics as incomprehensible as Latin for most Catholics.

---

[72] RUSSELL W. HOWELL and W. JAMES BRADLEY (eds.), Mathematics in a Postmodern Age: A Christian Perspective, Wm. Eerdmans Publishing Co., 2001; JOHN BYL, The Divine Challenge: on Matter, Mind, Math and Meaning, Banner of Truth Trust, 2004.

A reasonable religion that has the ambition to provide answers to those peoples' metaphysical questions, must take into account the scientific insights of the nature of the reality. Metaphysics refer to views about the world as a whole and religious communities use theological symbols to explain how the All reveals itself to human beings.

There is a paradoxical link between mathematics, reason and metaphysics. On the one hand, mathematical logic and metaphysics are two opposite dimensions of human knowledge with considerable differences between the two. On the other hand, there are some deeply rooted similarities between metaphysics and the pure deductive reason and mathematical logic is the area of human knowledge with a clearest use of pure deductive reason.

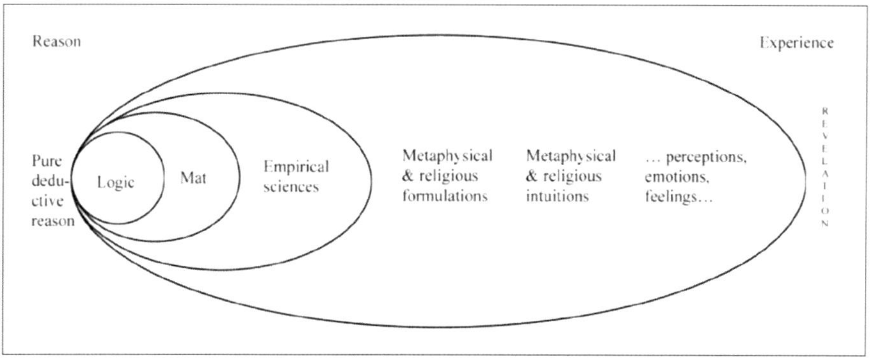

Differences between mathematics and metaphysics;

1. Signs vs. Symbols: Mathematical logic employs signs with an objective and defined meaning. Metaphysical propositions use symbols whose meaning refers to reality as a whole and often to a transcendent God.

2. Conveyable vs. Indescribable: Whereas the propositions in mathematical logic can be translated into all natural languages and can be conveyed to anyone, many mystics have expressed their linguistic inability to convey their

metaphysical experience about the ultimate sense and meaning of the world and also their relationship with God.

3. Contemplation vs. Control: Formal knowledge of the laws of nature allows reality to be controlled from a technological point of view. In metaphysical and religious views, understanding rests on a contemplative sense of mystery.
4. Definition vs. Holism: Mathematical and logical statements refer to a particular area of discussion. Metaphysical and religious propositions refer to reality as a whole.
5. Complexity vs. Simplicity: Knowledge of mathematical logic is specific, analytical and complex. Global metaphysical knowledge is man-made and simple.

Paradoxically, pure deductive reason is:

1. Simple, because it does not change
2. Holistic, because deductive reasoning is applicable to any kind of knowledge and is always the same in any place and at any time
3. Contemplative, because it does not depend on human activity.
4. Indescribable, because it is common to all languages.

Reason is actively present; structuring, organizing and clarifying etc. Religion is present in a more contemplative way that could be called metaphysical. Metaphysics is understood to mean a radical and basic view of the world where everything matters; everything is included, looking for a rational answer to the question about reality and realities as a whole.[73]

Structuring philosophical and religious formulations by pure deductive reason is the most solid foundation and point of contact not only for inter-disciplinary exchange between scientific communities and inter-cultural exchange between different human communities, but also for inter-religious exchange between different

---

[73] J. LEACH, MATHEMATICS, REASON & RELIGION, PENSAMIENTO, vol. 64 (2008), núm. 242.

religious communities. Metaphysical and theological arguments need to be able to span different cultural communities.

There is a need for a new metaphysic that is open to questions about the meaning of life, the meaning of existence and the universe. Symbolic anthropology and some versions of phenomenology argue that all humans require reassurance that the world is safe and ordered place – that is, they have a need for ontological security[74].

Therefore, all societies have forms of knowledge that perform this psychological task. The inability of science to offer psychological and emotional comfort explains the presence and influence of non-scientific knowledge in human lives, even in a rational world. The better people understand what moves things around them, the better they can position themselves to take advantage of these movements for personal growth and, subsequently, act in a way that makes the world a better place to live in for everyone.

It's very doubtful that the cosmic horizon, who's by science perceived as the originator of our reality, is receptive for prayers. But a rational religion could explain the way it communicates with us; synchronic events, climate change, epidemics ... are all examples of the All to address manmade disturbances of a natural balance.

Three values are the core of what people are looking for in a religious environment: friendship, love and self-esteem and -development. They look for truth, honesty and warmth. They don't like moralizers but are sensible for living testimonies. Important are small scale projects, human contact, space and group binding to share experiences and to know themselves sheltered, to come to rest and to revitalize themselves.

---

[74] Giddens, Anthony (1991). Modernity and self-identity: self and society in the late modern age, Cambridge, Polity Press.

Part 5; Towards a new synthesis.

5.1. Introduction

Our current understanding of the beginning of the universe is that it bubbled up from a pre-universal singularity called the homogametic principle.
We postulate that time and space started with the Big Bang that created the universe as we know it. Previous times and places are non-existent.
The reality as we know consists of informational modulated energy waves, emitted by a cosmical horizon upon the Higgs-field that is formed by the time-space continuum. The Schrödinger equation describes this wave – particle duality and its progress through the time-space continuum.
The geosphere and biosphere were the creation phases that were leading to intelligent self-conscious life that created a noosphere.
Since language is the most important tool that we use to transmit information, is it impossible for an individual to give an impartial description that supersedes the limitations of his language. The description of such information process can only through mathematization overcome such language barriers. Nevertheless, even then the universe will remain something we invent.
Since the latest evolutionary phase of the earth, the formation of a noosphere, is a mental construct that budded up from the human conscient thinking, we might assume that it is also ruled by some of the principles that govern human neurological processes.
One of the most fundamental principles that govern the human esthetical sense is the Fibonacci sequence. Citi planners should keep this in mind when they want to create an environment that is esthetically pleasing its inhabitants.
Our actual understanding of the neurological processes that lead to rational thought is that those processes are guided by the Bellman optimization equation.
This noosphere is the determining factor that causes that rise and fall of the different civilizations that each on their turn function as an

envelope of the progress and evolution of human knowledge. It seems to be a law of nature that the faster a civilization reaches its peak, the faster it declines. The mathematical model of human evolution and history has presented this in a graphical illustration of sequential exponentially growing curves.

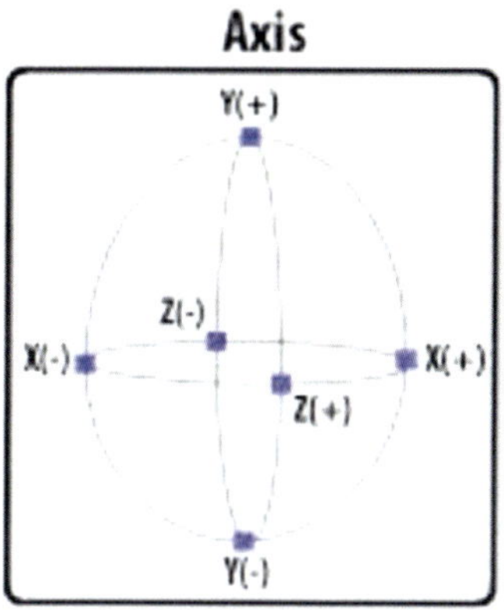

Each envelope could also be described as an egg that contains all the knowledge that humanhood has accumulated over time. This egg is made by rotating the curve of the evolutionary model around the peak axis.

In this simplified model the point y(-) symbolizes the birth of a new civilization that absorbs the knowledge if the previous model till it reaches points x(-), z(-) and z(+), which are respectively evolutions in science, art and religion till all three of them fusion at the peak of the civilization in point y(+).

From there on the perpendicular sections diminish in size till they are reduced to zero at point x(+).

It can be safely assumed that each civilization thus created its own variation of the "Philosopher´s Egg".

5.2. The US as the dominant envelope.

With hindsight we can easily state that the US as an envelope for the noosphere reached its peak around 1965. It was a time of growing secularization and emphasis on human rights on the religious field, coinciding with the moon landings and the science that made this possible. Einstein and Alan Turing were undoubtedly the most significant scientists of the 20th century. The artistic scene was blossoming, with artists as Bob Dylan, Andy Warhol and Harper Lee, their art being the ultimate expression of an industrialized and secularized society. The most prominent religious leader of that time

was dominee Martin Luther King who was putting human rights ahead in all his preaching.
The US is the epitome of the second wave nation that gave birth to a third wave revolution but tries to handle the transit towards an information society with a second wave mentality and methodology.
After 1965 we see a diminishing interest of US citizens in art, as illustrated by the almost flatlining subject cycle of the bestselling novels. On a religious level there is a return towards (mostly Biblical) mysticism and an increasing degree of human rights violations. Science is fundamentally a spinoff from the space technology that put the first man on the Moon. Most higher education programs are reduced to a "teach the test" level, while the more ambitious programs are inaccessible for most of its population. They tend to overspecialize in a very narrow field of interest with neglect for the bigger picture. Nobody is really interested in a discussion similar as "How many angels can dance on the top of a needle?"
To put it more succinctly: the US civilization is in its "Bread and Games" phase. It will continue to wield a considerable influence on human society for the coming decennia, but this rests upon realizations made into the past while there is a increasing social entropy growing inside this nation-system.

5.3 The Upcoming envelopes.

The upcoming evolutionary envelope, China, is officially an agnostic nation although a big amount of its citizens are practicing Taoists. This is a metaphysical belief that divides the forces ruling our universe in Yin and Yang. Order and chaos are the two elements that give shape to our reality and neither of them is fundamentally good nor bad since both aspects can be driven by rightful or evil intentions. One just must think about the absolute order that most dictatorial regimes are craving to implement or the lawless opportunism that rules in a so called "failed" state. This coincides much with recent insights into the system theory, the lens through which scientists start to interpret our reality.

The Chinese system of Fengshui is finely tunes with our esthetic sense that is (at least partially) based upon the recognition of Fibonacci sequences. As most upcoming envelopes, it is characterized by an authoritarian leadership with uttermost neglect of human rights. This stance of the Chinese leadership encounters a growing degree of disapproval among its citizens, fueled by nepotism and corruption among the ruling elite. This dissidence finds its expression in many artistic works, an activity that is heavily censored but widely shared among the population through the omnipresent information technology.
Singapore is leading the way among the Asian tigers as the leading financial center of Asia but is because of its geographical location and small size very vulnerable for expansionist tendencies of neighboring countries. Its education system and healthcare are topnotch and can easily withstand any comparison with most Western institutions. The Singaporean leadership practices an enlightened despotism mixed with some democratic elements with rules that govern almost all facets of its inhabitants. By most visitors it is perceived as a very repressive law and order society.
It has no doubt that a regime that wants to exercise an absolute power over its population is slowing down its society´s progress. A civilization can only peak when Bellman´s optimization theorem is applied to all facets of a civilization.

.

5.4. Will Bhutan become the new Switzerland of the Himalaya´s?

To me it seems that the example that has been set down by Bhutan indicates the future evolution of our model for a human society that can implement the new insights of our information age and the values of an ecological and decentralized socio-economical model that places the interests of the individual balanced against the needs of society in the center of its considerations. This nation adheres almost homogenic to the Buddhist religion.

This religion is also spreading among digital nomads. Widespread musical and artistic experimentation; the emergence of 'youth culture' as a powerful economic force; the increasing availability of mind-altering recreational drugs; the collapse of western colonial projects; and the intense controversies around the American Civil Rights movements and the wars in Vietnam and the Middle East, all contributed to a profound questioning of received values.

Cheap international travel and the diaspora of Asian Buddhist teachers in the West enabled people to experiment with practical methods such as meditation and chanting. These practices contrasted with the religious beliefs and rites that many had grown up with, and that could be deemed passive and stuffy by skeptics.

People come to Buddhism expecting something different. They often have serious spiritual or life questions in mind, but at the same time some are fed up with what they see as conventional religion.

In a similar vein, the absence of elaborate hierarchies or involvement in politics makes Buddhism attractive in our egalitarian age. And as established religion has diminished as a source for authoritative guidance about how to live and what to live for, many people have sought fulfillment and understanding in a more personal way from science and the arts.

In particular, the religions familiar in the West, Christianity, Judaism and Islam, all situate humanity in a finite timeline, running from revelation to an 'end time' - all of which is being managed by God, whereas Buddhism places the human situation in an infinite and neutral cyclical cosmos. Some people find this more compatible with their understanding of the world.

Other people have explicit problems with "God". Buddhism side steps God and offers a vision of humanity and the spiritual life that is not dependent on a deity. It lays out a path of self-transformation, and people like the idea of being in control and taking responsibility for their lives that can come with this.

Looked at another way, Buddhism offers seemingly practical ways of developing wisdom and compassion, free from guilt and obligation. And of course, I think there is the simplest explanation of all for why Buddhism is hip, which is that it offers something that is helpful and meaningful, and that people just like it for what it is in itself.

The Bhutanese metaphysical system is apt to absorb and grow along the lines that are put forward by the emerging new values. However, there are some factors that keep Bhutan from developing itself into the next envelope of human civilization. They are:

1. Isolationist tendencies partially induced by its geography and a fear for losing their cultural identity and independency to more powerful neighbors. The Bhutanese obsession with shielding their own culture against outside influences leads sometimes to xenophobic reactions. Most of their artistic and intellectual activity is currently oriented towards preservation efforts of their mostly oral and artisanal traditions. This leaves little space for new ideas to perspire into the artistic and intellectual field that are not implemented from the top towards the bottom.
2. The country still battles with a staggering illiteracy rate of 40 % and the higher education level is still sub-standard.
3. Bhutan with its about 1 million habitants carries almost no social weight in the international environment. Its population remains far below the critical mass needed for a further propagation of their model of society. One could consider it as a social laboratory where new ideas are tested.
4. Another crippling factor is the absence of an industrial infrastructure that causes a dependency on import of high technological hardware needed for developing the desired sociological model (the country made a direct transition from an agrarian society towards an information-oriented culture).

Bhutan has the potential to develop into a New Age clone of Switzerland among the Asian tigers that surrounds it. But it still has a long way ahead to reach such goal.

Appendix A; Mathematical proof for the use of Fournier's Theorem to study periodic phenomena.

By the help of Fourier's analysis, a periodic function may be put in the form

$$(1)\ y = A_0 + a_1 \cos kt + a_2 \cos 2\,kt + a_3 \cos 3\,kt + \ldots$$
$$+ b_1 \sin kt + b_2 \sin 2\,kt + b_3 \sin 3\,kt + \ldots$$

If in (1), we put

$$a_1 = A_1 \sin e_1;\ a_2 = A_2 \sin e_2;\ a_3 = A_3 \sin e_3;\ \&c.,$$
$$b_1 = A_1 \cos e_1;\ b_2 = A_2 \cos e_2;\ b_3 = A_3 \cos e_3;\ \&c.,$$

We get

$$(2)\ y = A_0 + A_1 \sin (kt + e_1) + A_2 \sin (2\,kt + e_2)$$
$$+ A_3 \sin (3\,kt + e_3) + \ldots$$

where y is expressed as a series of sinuses. In a similar manner, equation (1) may be expressed as a series of cosines,

$$(3)\ y = A_0 + B_1 \cos (kt - \epsilon_1) + B_2 \cos (2\,kt - \epsilon_2)$$
$$+ B_3 \cos (3\,kt - \epsilon_3) + \ldots$$

In the use of Fourier's theorem for the purpose of analyzing periodic phenomena, we follow a process analogous to the use of Taylor's theorem in the simpler demonstrations of mathematical economics. On the assumption that, if the literary function under investigation is y = f (x), then f (x+h) may be expanded by Taylor's theorem[75], and the first terms of the series may be used as an approximation to the

[75] Taylor's theorem states that any function satisfying certain conditions may be represented by a Taylor series. Jeffreys, H. and Jeffreys, B. S. "Taylor's Theorem." §1.133 in Methods of Mathematical Physics, 3rd ed. Cambridge, England: Cambridge University Press, pp. 50-51, 1988.

form of f(x). Similarly, in our use of Fourier's series, the attention will be focused upon a few harmonics as a first approximation to the solution of the problem in hand of expressing in mathematical form the periodicity of literary phenomena. Assuming that any periodic function may be expressed as a Fourier series, the problem is presented of determining the values of the coefficients. The series, as we know, is of the form

$$y = f(t) = A_0 + a_1 \cos kt + a_2 \cos 2\,kt + \ldots$$
$$+ b_1 \sin kt + b_2 \sin 2\,kt + \ldots$$

What are the values of the first term and of the coefficients of the sines and cosines? In order to deduce the necessary values, we shall have need of the following lemma:
If m and n are two unequal integers and k is put equal to

$\frac{2\pi}{T}$, then

$$\int_0^T \cos mkt \cos nkt \, dt = 0,$$
$$\int_0^T \sin mkt \sin nkt \, dt = 0,$$
$$\int_0^T \sin mkt \cos nkt \, dt = 0.$$

The lemma may be proved to be true by evaluating the three integrals according to the usual methods. The first integral, for example, becomes

$$\int_0^T \cos mkt \cos nkt \, dt = \tfrac{1}{2} \int_0^T \{\cos (m-n)\, kt + \cos (m+n)\, kt\}\, dt$$
$$= \left[\frac{\sin (m-n)\, kt}{2\,(m-n)\, k} + \frac{\sin (m+n)\, kt}{2\,(m+n)\, k} \right]_0^T$$

But $k = \frac{2\pi}{T}$, and, consequently, $\int_0^T \cos mkt \cos nkt \, dt = 0.$

With the aid of this lemma we may proceed to evaluate the coefficients in Fourier's series. If we integrate the series between the limits o and T, we get,

$$\int_0^T f(t)\,dt = A_0 \int_0^T dt + a_1 \int_0^T \cos kt\,dt + b_1 \int_0^T \sin kt\,dt + \ldots$$

But all of the terms except the first on the right-hand side of the equation will vanish, and consequently

$$\int_0^T f(t)\,dt = A_0 \int_0^T dt = A_0 T, \text{ or } A_0 = \frac{\int_0^T f(t)\,dt}{T}$$

Since

$$\int_0^T f(t)\,dt$$

is the area of the original curve for one whole period T, the constant term in Fourier's series is equal to the value of the mean ordinate of the original curve.

To determine the value of all multiply throughout by ***cos kt*** and integrate between limits O and T.

$$\int_0^T f(t)\cos kt\,dt = A_0 \int_0^T \cos kt\,dt + a_1 \int_0^T \cos^2 kt\,dt$$
$$+ b_1 \int_0^T \sin kt \cos kt\,dt + \ldots$$

or

$$\int_0^T f(t)\cos kt\,dt = a_1 \int_0^T \cos^2 kt\,dt,$$

since

$$\int_0^T \cos kt\,dt$$

and

$$\int_0^T \sin kt \cos kt\,dt$$

are both equal to zero and all the other terms on the right-hand side of the equation, according to our lemma, disappear. But

$$\int_0^T \cos^2 kt\,dt = \int_0^T \frac{1 + \cos 2kt}{2}\,dt = \tfrac{1}{2}\left[t + \frac{\sin 2kt}{2k}\right]_0^T = \frac{T}{2}$$

and as a result, we have

$$a_1 \frac{T}{2} = \int_0^T f(t) \cos kt \, dt, \text{ or } a_1 = 2 \frac{\int_0^T f(t) \cos kt \, dt}{T}.$$

Therefore al is equal to twice the mean value of the product ***f(t)cos kt***. In a similar manner the value of any other coefficient may be determined. Take, for example b_n . Multiply throughout by ***sin nlct*** and integrate between ***o*** and ***T***,

$$\int_0^T f(t) \sin nkt \, dt = b_n \int_0^T \sin^2 nkt \, dt = b_n \int_0^T \frac{1-\cos 2\,nkt}{2} dt =$$

$$b_n \left\{ \tfrac{1}{2}\left[t - \frac{\sin 2\,nkt}{2\,nk} \right]_0^T \right\} = b_n \frac{T}{2}$$

and, consequently

$$b_n = 2 \frac{\int_0^T f(t) \sin nkt \, dt}{T}.$$

therefore b_n is equal to twice the mean value of the product ***f(t) sin (nkt)***.

Appendix B; Arms races

Mathematically, mutually reinforcing hostility can be expressed as a pair of differential equations. Let x represent the "amount of hostility" expressed by X toward Y. (For the moment, we by-pass the question of how this "amount of hostility" can be measured.) Y's level of hostility, y, will now be assumed to be proportional to the rate of change of X's hostility, and vice versa. We write

$$\frac{dx}{dt} = ay \qquad (a > 0) \tag{3.14}$$

$$\frac{dy}{dy} = bx \qquad (b > 0). \tag{3.15}$$

Differentiating Equation 3.14 with respect to t, we have

$$\frac{d^2x}{dt^2} = a\,\frac{dy}{dt}. \tag{3.16}$$

Substituting into the right side of Equation 3.15, we obtain

$$\frac{d^2x}{dt^2} = abx. \tag{3.17}$$

The general solution of Fquation 3.17 is

$$x(t) = Ae^{\beta t} + Be^{-\beta t}, \tag{3.18}$$

where $\beta = \sqrt{ab}$ and A and B are determined by the initial conditions $x(0)$ and $y(0)$, whereby $y(t)$ is also determined. Since $\beta = \sqrt{ab} > 0$, $x(t)$ increases without bound as t tends to infinity, and so does $y(t)$. Thus the model loses its physical significance. As in the case of population growth, it is necessary to introduce constraints. These enter the differential equations as negative terms on the right side.

$$\frac{dx}{dt} = ay - mx \tag{3.19}$$

$$\frac{dy}{dt} = bx - ny. \tag{3.20}$$

To make use of these equations in a predictive theory, the variables must designate some observable quantities. Thus we need a concrete index of the "amount of hostility." Richardson chose the armament budgets of nation states as an index of perceived hostility in line with his argument that the level of armaments of a state (or a bloc of states) is perceived as evidence of hostile intentions by another state or bloc that feels threatened by the other's war preparations.

Appendix C; Levitation forces induced by superconductors

The levitation force

According to the similar discussion as before [9], the top surface center of the SD approaches the bottom surface center of PM as,

$$s = z_0 - vt \tag{11}$$

where velocity v represents the speed at which the PM approaches the SD, z_0 is the initial distance between the top surface of the SD and the bottom surface of the PM.

As the current density $J(\rho, z, t)$ and the radial magnetic field B_ρ^{PM} inside the SD have been derived, the vertical levitation force along the z-axis can be readily obtained as [9],

$$F_z = 2\pi \int_0^a \rho d\rho \int_{-b}^{b} dz J(\rho, z) B_\rho^{PM}(\rho, z) \tag{12}$$

Selection of Parameters

The parameters used in our calculation are taken as follows: $T_{c0} = 92$ K [9]; $\mu_0 = 4\pi \times 10^{-7}$; $C = 0.88 \times 10^6$ J/m^{3}K [8]; $\kappa = 6$ W·m^{-1}·K^{-1} [8]; $E_c = 1\times10^{-4}$ V/m [10]; $\alpha = 1.9\times10^9$ A/m^2; $v = 2$ mm/s; $B_0 = 0.5$T and the flux creep exponent $\sigma = 4$. All these parameters remain unchanged

Appendix D; Technical elaboration upon the proposed mathematical neurological model

Optimal decision making in a complex world is a challenging computational task, without constraining computations to be performed in a biologically plausible manner, which was the problem that was addressed here. Neural network instantiations had been suggested for model-free reinforcement learning using gradient-based methods as well as temporal-difference methods.

However, this model is the first biologically realistic implementation for complex model-based decisions. For biological plausibility, it used a canonical, phenomenological single-neuron model, the spike–response model (which gives a simple description of action potential generation in neurons.).

The spike response model is the most general model that combines linear filtering with a strict threshold. Hence it incorporates several spiking neuron models as special cases or, equivalently, the generalized linear model.

The generalized linear model (GLM) is a flexible generalization of ordinary linear regression that allows for response variables that have error distribution models other than a normal distribution and consists of three components:

- A random component,
- A systematic component, and.
- A link function

It has been demonstrated to be an accurate predictor of neural responses in a wide variety of brain areas and a synaptic plasticity rule that was local and Hebbian in nature.

Hebbian theory is a neuroscientific theory claiming that an increase in synaptic efficacy arises from a presynaptic cell's repeated and persistent stimulation of a postsynaptic cell. It is an attempt to explain synaptic plasticity, the adaptation of brain neurons during the learning process.

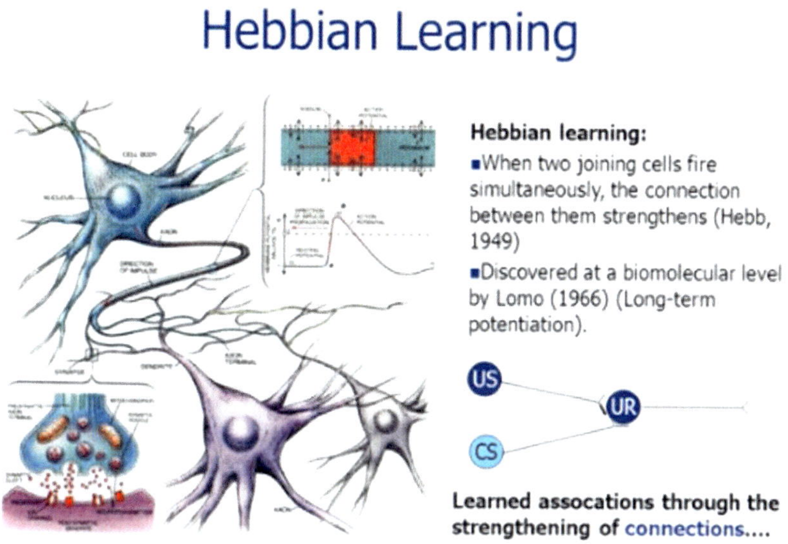

Hebbian Learning
Hebbian learning:
▪When two joining cells fire simultaneously, the connection between them strengthens (Hebb, 1949)
▪Discovered at a biomolecular level by Lomo (1966) (Long-term potentiation).
US
UR
CS
Learned assocations through the strengthening of connections....

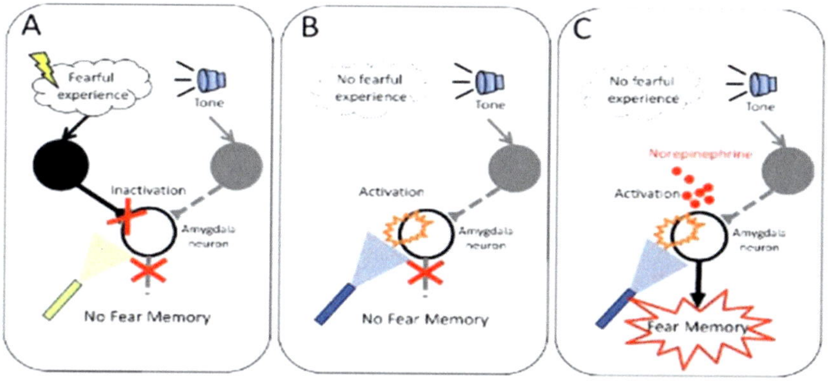

It was based on prediction errors to acquire an appropriate internal model of the environment. It showed that model circuits constructed from such elements achieved competent performance in model-based sequential decision making and that the neural dynamics predicted by the model are consistent with a broad range of experimental data.

The authors suggest that their network resides in the PFC (pre-frontal cortex).

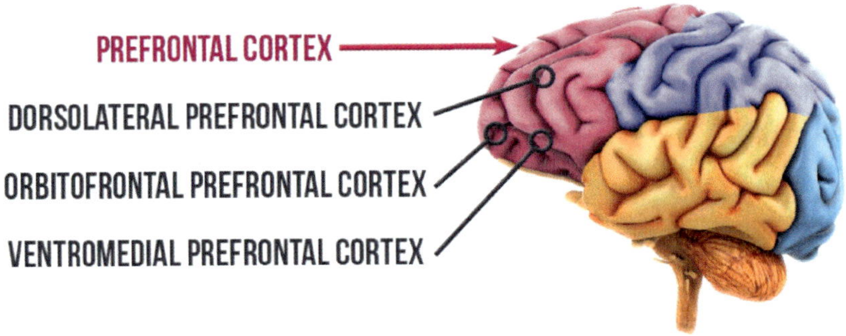

More specifically, they hypothesize that the state–action neurons, of which the dynamics our network explicitly modeled, correspond to so-called offer value cells found in Orbital FC. Whether the NRM-modulated neurons of pre-SMA receive inputs from these neurons or implement a parallel network is unclear, but our model suggests that they should be similarly modulated by values as OFC neurons are and, conversely, OFC neurons should show similar activation patterns as pre-SMA neurons in sequential decision-making tasks. They further hypothesize that the accumulation of spike count differences underlying the final decision takes place in ventromedial prefrontal cortex, which receives input from OFC and has been implicated in the comparison of options.

This network was constructed such that neurons encoded (or preferred) specific state–action pairs. However, empirical data suggest that actions may be represented in an ordinal fashion, such that even the same action in the same state may be represented by different neurons depending on how distal (i.e., how many time steps away) it is to the current state of the animal. It is straightforward to extend this model to represent state–action–time step tuples by replicating the state–action neurons for each time step in our current network and having excitatory synapses connect to neurons encoding the previous time step. Whereas such an implementation allows for time-dependent policies, it deals only with the case of finite horizon and comes at the expense of growing the size of the network with time horizon. Nevertheless, based on such a representation, action–time step neurons can be obtained by projecting down into a downstream area, akin to obtaining chosen value cells from offer value cells in OFC, and thus account for ordinal action representations. To make this model metabolically more efficient, and more comparable in that regard to earlier models, it could further be extended such that a separate network computing the reachability of future states from the current one (e.g., by forward spread of activation) provides an extra input to our network so that only neurons representing reachable states (and actions) in it are close to the threshold.

To simplify exposition, the network is presented as a minimal neural circuit allowing neurons to have both excitatory and inhibitory synapses. The violation of Dale's principle (which basically states that a neuron performs the same chemical action at all of its synaptic connections to other cells, regardless of the identity of the target cell) could be avoided by explicitly considering the interneurons[76] that mediate lateral inhibition between all excitatory cells coding for the same state (but potentially different actions).

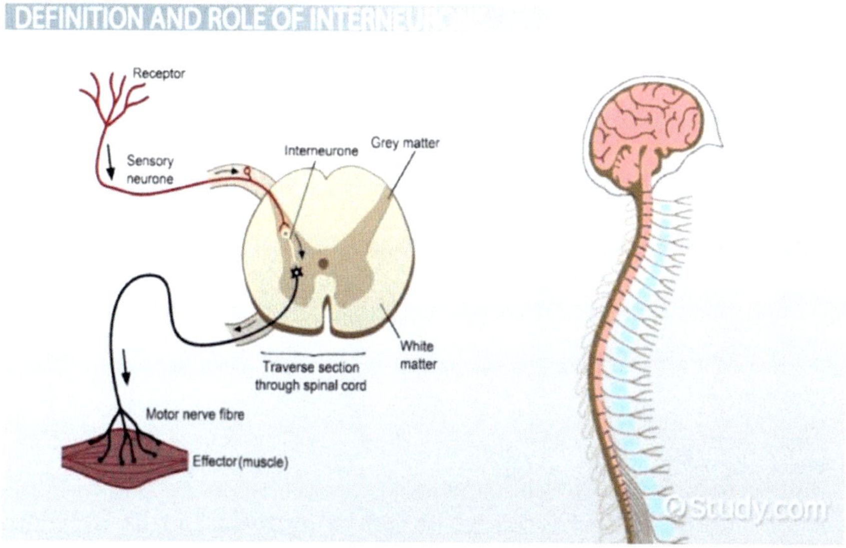

Moreover, disynaptic inhibition (an effect from the one cell to another via two synapses, the response depending on the rate and duration of the presynaptic activity) may also account for offer value

[76] Interneurons are the central nodes of neural circuits, enabling communication between sensory or motor neurons and the central nervous system (CNS). They play vital roles in reflexes, neuronal oscillations, and neurogenesis in the adult mammalian brain

cells showing negative modulation by value and pre-Secondary Motor Area cells negatively modulated by the superior central nucleus, which this simplified network could not capture.

Lateral inhibition was a key element of the dynamics of this network.

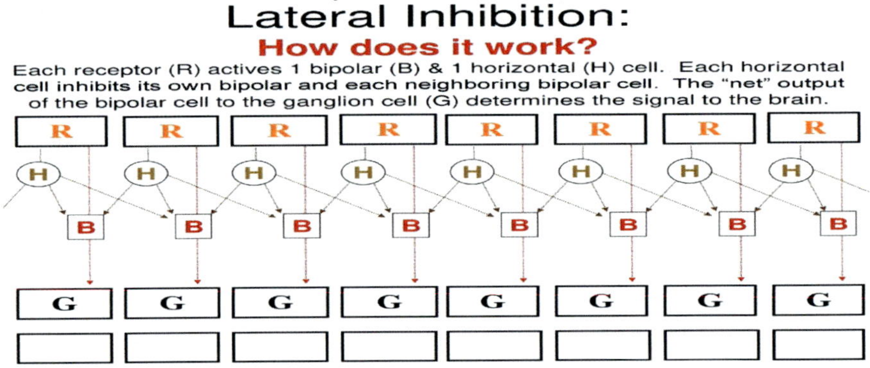

Lateral Inhibition:
How does it work?
Each receptor (R) actives 1 bipolar (B) & 1 horizontal (H) cell. Each horizontal cell inhibits its own bipolar and each neighboring bipolar cell. The "net" output of the bipolar cell to the ganglion cell (G) determines the signal to the brain.
R
H
B
G

Lateral inhibition is the capacity of an excited neuron to reduce the activity of its neighbors. Lateral inhibition disables the spreading of action potentials from excited neurons to neighboring neurons in the lateral direction. This creates a contrast in stimulation that allows increased sensory perception.

While many previous models used lateral inhibition to implement a winner-take-all mechanism between different choices, there are important differences between those models and this one. This is because the precise ways in which lateral inhibition acts in a model, and in particular whether it acts between units having neural-like or longer time constants, can have profound consequences for its dynamics.

Sequential decision making involves algorithms that take the dynamics of the world into consideration, thus delay parts of the problem until it must be solved. It can be described as a procedural approach to decision-making, or as a step by step decision theory. Sequential decision making has as a consequence the intertemporal choice problem, where earlier decisions influence the later available choices.

Models of nonsequential perceptual decision making typically use long (~100 ms) time constants that match the time scale of individual trials to achieve reliable accumulation of evidence, whereas this network uses shorter, more realistic membrane time constants (~20 ms) and thus relied on accumulation happening as a "postprocessing step." This difference has important implications for the interplay between accumulation and lateral inhibition.

In neurobiology, lateral inhibition is the capacity of an excited neuron to reduce the activity of its neighbors. Lateral inhibition disables the spreading of action potentials from excited neurons to neighboring neurons in the lateral direction.

In earlier models, lateral inhibition occurred at the level of the accumulated decision variables.

In contrast, in this model, which might otherwise appear a close (spiking) analog of these earlier models in the case of two-alternative forced choices, it affected already the nonaccumulated decision variables represented by the neurons of this network. Inhibition between accumulated variables tends to lead to unrealistically skewed reaction time distributions, which this model avoided but which are also found in simulations of an alternative variant of this network lacking inhibition between fast time scale neurons (data not shown). Moreover, the specific form of lateral inhibition used in this network was also crucial to allow to generalize it to solving the ecologically more relevant and computationally more challenging task of multistep sequential decision making (in contrast to simple spreading activation models that do not use lateral inhibition; see figure next page) and also predicted neural and behavioral data in such richer tasks with high accuracy.

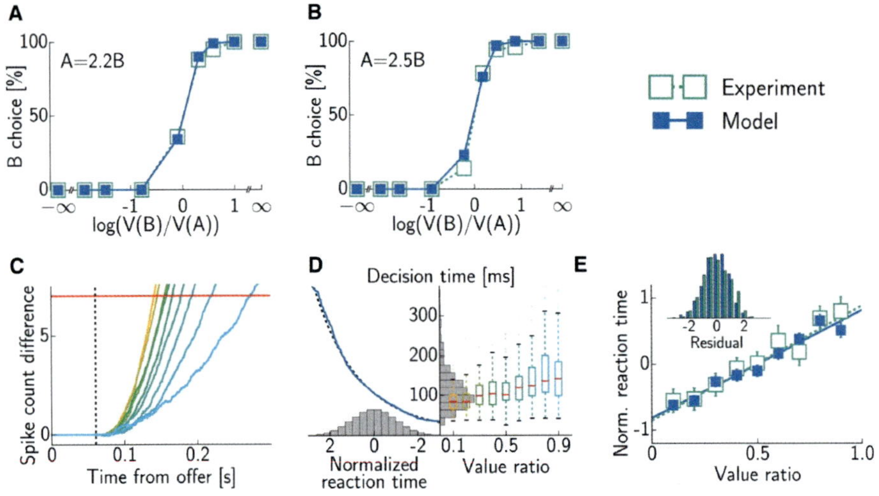

Psychometric and chronometric curves in a binary decision-making task. A, B, Choice probabilities in experiments (open green squares; Padoa-Schioppa and Assad, 2006) and simulations (filled blue squares) for two different relative values of the two juices: 1A = 2.2B (A) and 1A = 2.5B (B). C, Difference between the cumulative spike counts of populations representing the two potential choices in the model. Accumulation starts with sensory delay (dashed line; compare input onset in Fig. 3B). When a threshold (red line) is reached, a decision is made. Colors indicate different value ratios as in D. D, Decision time distributions in the model. Right, Dependence of raw decision times on the value ratio (colored Tukey's boxplots) and their overall distribution across all value ratios (gray histogram). Left, Normalizing function (solid blue line), together with a logarithmic fit (dashed black line), which transforms the raw decision time distribution into a standard normal distribution (gray histogram). E, Normalized reaction times (±SEM) as a function of value ratio in experiments (open green squares; Padoa-Schioppa and Assad, 2006) and simulations (filled blue squares). Lines show least squares fits (dotted green, experiments; solid blue, simulations); the inset shows distribution of residuals after fitting (green bars, experiments; blue bars, simulations).

As a consequence of the specific role that lateral inhibition plays in its dynamics, this model also provides an alternative account of the stochasticity of decisions and the distribution of decision times in simple perceptual decision-making tasks in which the stimulus is not explicitly stochastic. Previous models relied on stochasticity in the inputs to the network, even for nonstochastic stimuli, and the fact that this stochasticity needs to be integrated out over time. Thus, both psychometric and chronometric curves primarily depended on this external noise. In contrast, this network receives deterministic input, and so psychometric curves are a result of the spiking "noise" within the network itself, whereas chronometric curves are a consequence of the specific form of lateral inhibition used in it.

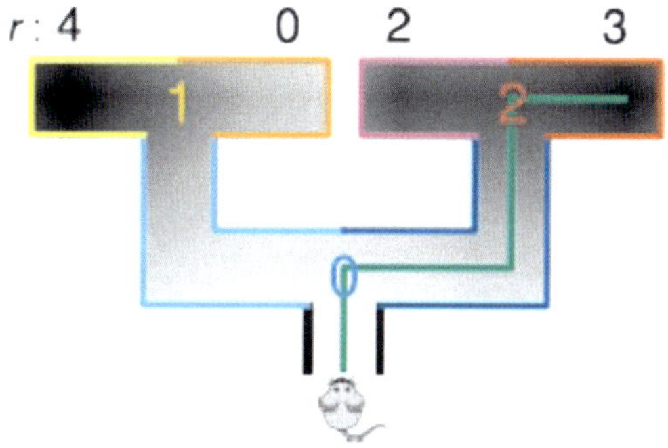

Activity profile for a spreading-activation model (darker means increasing activity). The path of an agent following the activity gradient (green) yields only a reward of 3 instead of the optimal 4.

This approach to perform sequential decision making was based on the principle of dynamic programming: instead of performing a sequential tree search, planning in this model occurred in a near-parallel fashion, as suggested by near-parallel neural activations observed in PFC. Importantly, dynamic programming-based, goal-directed decision-making algorithms, such as that implemented by our network, require an internal model of task contingencies that needs to be acquired through interactions and experience with the environment. For this, this network required that the same neurons that encode states and actions during planning of an action sequence become activated later while that sequence is being performed, so that the synaptic weights between neurons faithfully reflect the transition and reward probabilities implied by the task. Such reactivation of neurons taking part in planning and execution has also been observed in the PFC.

Off-line replay of experience, during periods of rest or sleep, as observed throughout the neocortex and, in particular, in multiple brain areas implicated in goal-directed decision making, such as the ventral striatum, the PFC, and the hippocampus, may also serve to reinforce and consolidate internal models of the environment

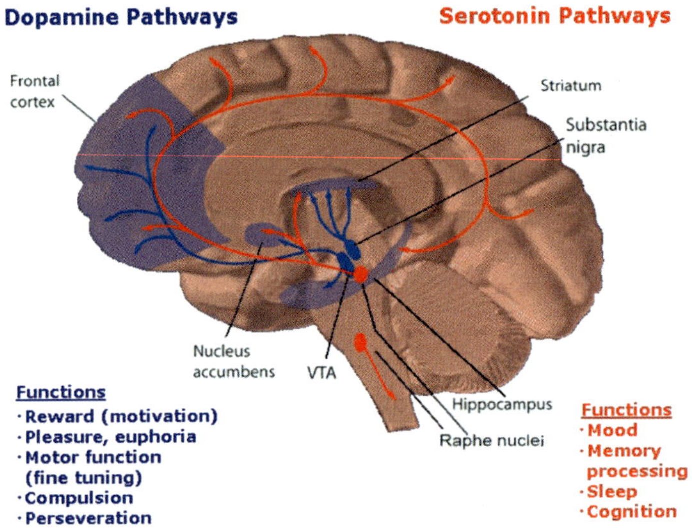

Although several other models have been proposed in which the replay of experience underlies model learning, this model differs from these in a crucial aspect. Although those models used Hebbian plasticity to store information about the experienced sequences, this model requires either the replay of sequences in reverse temporal order or forward replay to be coupled with anti-Hebbian forms of plasticity (such that post-before-presynaptic activation is needed for potentiation). This is because in this model, an excitatory synaptic weight connecting a presynaptic neuron i to postsynaptic neuron j represents the probability of reaching the state represented by neuron i from that represented by neuron j. Reverse replay has only been observed in the hippocampus and it remains to be tested whether, for example, it also exists in PFC or whether there are anti-Hebbian forms of plasticity operating there.

On-line reverse hippocampal replay in the form of spreading activation has also been suggested to provide the neural substrate of spatial navigation. Considering navigation a special case of sequential decision making allows a direct comparison between these models and the one at hand. In line with previous work, this model predicts activity spreading from neurons representing the goal to neurons representing more proximal locations. However, the precise form of spreading activation in this network is different, and notably nonlinear, thus allowing the network to solve the more general problem of maximizing return in sequential decision-making tasks with multiple rewards, for which classical spreading activation models would fail to account.

Therefore, this model also makes the novel prediction that the replay-like phenomenon of spreading activation during planning should also generalize to distinctly nonspatial domains (albeit perhaps in other cortical areas).

Previous proposals for how cortical circuits may solve sequential decision tasks were based on the powerful idea of using probabilistic inference algorithms for planning. Although this idea is conceptually and algorithmically attractive, especially in light of the converging evidence that cortical circuits may naturally perform probabilistic inference, the neural instantiations suggested so far relied on two particularly speculative assumptions: they required multiplicative interactions between presynaptic neurons and assumed that dendrites approximately perform a logarithmic transformation on their inputs. Furthermore, state and action neurons needed to be replicated for each time step of a sequential task; hence, the size of the network grew with the time horizon, which had to be finite. Moreover, as an illustrative example task revealed (see figure below), the particular form of probabilistic inference afforded in these networks (belief propagation) leads to severely suboptimal behavior in even simple test cases, which this network successfully solved. Whereas belief propagation has been often evoked as the algorithmic basis of how the cortex performs probabilistic inference, recent results suggest that the cortex may, in fact, implement other kinds of inference algorithms that are rich enough to capture the structure of any decision-making task. Thus, even simple tasks such as those presented here could be used to discriminate competing proposals for the type of algorithm that the cortex implements for model-based decision making.

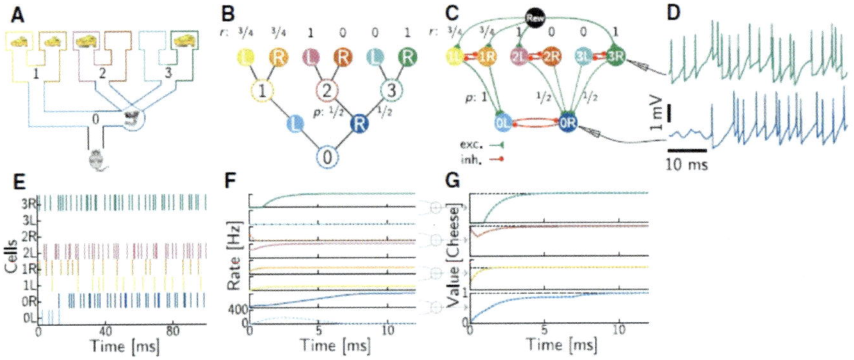

Two-step example task. A, The rat moving through the maze can choose the left (L) or right (R) arm at four decision points (states 0, 1, 2, and 3). Turning right in the first step (state 0) leads to a place where one of two doors opens randomly, indicated by the coin flip. The sizes of the cheeses indicate reward magnitudes (see also B). B, The decision graph corresponding to the task in A is a tree for this task. Numerical values indicate rewards (r) and transition probabilities (p) for nondeterministic actions. C, The corresponding neural network: action nodes in B are identified with neurons (colors). Lines indicate synaptic connections, with thickness and size scaled according to their strength. A constant external input (black) signals immediate reward. Synaptic efficacies are proportional to the transition probabilities or the (expected) reward. D, Voltage traces for two neurons in C. E, Spike trains of all neurons. The color code is the same as in C. F, Activity for rate neurons with random initial values. The color code is the same as in C. The line style indicates neurons coding for optimal (solid) and suboptimal (dashed) actions. G, The approximate values $\tilde{V}$, represented by the sum of the rates in F, converge to the optimal values (black dashed lines). Values of states 0–3 are shown from the bottom to top. The color code is the same as in B

TABLE I: Best sold literature 1561 - 1862

1561	Institutes of Christian Religion	Calvin	Theology	230.42
1563	Book of Martyrs	Foxe	martyrs	272
1605	Advancement of Learning	Bacon	Science--Methodology	192.1
1611	Bible		Bible.--Psalms	264.03
1612	A Map of Virginia with a discription of the Country	Smith	Virginia	917.55031
1616	A Description of New England: or the observations and discove	Smith	New England	973.2
1625	Essays	Bacon	English essays--Early modern	824
1640	Bay Psalm Book	Mather	Bible.--Psalms	264.03
1662	The Day of Doom	Wiggleswor	Judgment Day	281.4
1663	Hudibras	Butler	English poetry--Early modern	827.4
1665	The Great Assize	Smith	Judgment Day	281.4
1669	No Cross, No Crown	Penn	Christian life	248.4896
1672	An alarm to unconvinced sinners	Alleine	conversion	243
1678	Pilgrim's progress	Bunyan	Christian pilgrims and pilgrimages	263.04246
1682	Narrative of the captivity of Mrs Mary Rowlandson	Rowlandsor	Indian captivities	973.24
1683	The New England primer	Harris	congregational churches	428.6
1690	Essay concerning human understanding	Locke	Knowledge	121
1699	Journal; or God's protecting providence	Dickinson	Shipwreck survival	973.21
1702	The true history historical Narrative of the rebellion and Civil wa	Clarendon	Puritan Revolution (Great Britain : 1642-1660	942.062
1706	Horae lyricae	Watts	liturgic songs	245
1707	The redeemed captive returned to Zion	Williams	Queen Anne's War (United States : 1702-17	975.702
1709	The Tattler	Addison	Essays and belles lettres	824.52
1711	The spectator	Addison	English literature	824.5
1715	Divine songs for children	Watts	Christian education of children	245.21
1716	Iliad	Pope	Epic poetry	883
1719	Robinson Crusoe	Defoe	Shipwreck survival	973.21
1725	History of my own times	Burnet	Great Britain	942.06
1726	Odyssey	Pope	Epic poetry	883
1727	Astronomical diary and almanac	Ames	Astronomy	529.43
1728	Guliver's Travels	Swift	Shipwreck survival	973.21
1729	The seasons	Thomson	Seasons	525.5
1733	Poor Richard's almanac; poor Richard improved	Franklin	caricatures	741
1734	An essay on man	Pope	Knowledge	121
1740	Pamela	Richardson	Abduction	343.6
1742	The complaint, or night thoughts on life, death and immortality	Young	Knowledge	121
1743	The grave	Blair	Knowledge	121
1747	Clarissa Harlowe	Richardson	Abduction	343.6
1749	Tom Jones	Fielding	Illegitimate children	341.23
1751	Family companion or the oeconomy of human life	Dodsley	conduct of life	159.9
1757	The adventures of Peregrine Pickle : in which are included	Smollet	Tobacco	613.1
1760	The life and opinions of Tristram Shandy	Sterne	Hypocrisy	159.98
1761	Julia: or The new Eloisa	Rouseau	Man-woman relationships	159.96
1764	Castle of Otranto	Walpole	Inheritance and succession	347.6
1765	Commentaries of the laws of England	Blackstone	laws of England	349.42
1766	The vicar of Wakefield	Goldsmith	Abduction	343.6
1769	History of the reign of Charles V	Robertson	Europe--Holy Roman Empire	943.031
1770	The deserted village	Goldsmith	Extinct cities	973.3
1774	Letters	Chesterfield	Politics and government	321
1776	Common sense	Paine	conduct of life	159.9
1777	The American crisis	Paine	war	327.6
1778	M'Fingall	Trumbull	war	327.6
1783	American Spelling book	Webster	English language--Pronunciation	428.1
1787	The federalist papers	Hamilton	Constitutional law	342.4202
1791	Charlotte Temple; A tale of truth	Rowson	Illegitimate children	341.23
1792	Modern Chivalry	Backenridg	Politics and government	321
1792	Ruins	Volney	Revolutions	333
1794	Mysteries of Udolpho	Radcliffe	Inheritance and succession	347.6
1795	The age of reason	Paine	Theology	230.42
1796	The monk	Lewis	Man-woman relationships	159.96
1797	The coquette	Foster	Man-woman relationships	159.96
1799	Pleasures of hope	Campbell	Hope	152.52
1800	The life of Washington	Weems	Founding Fathers of the United States	973.3

1803	Thaddeus of Warsaw	Porter	Poland – History – Stanislaus II Augustus, 1	914.38
1805	The lay of the last minstrel	Scott	Fore-edge painting	96.1
1807	Hours of idleness	Byron	Man-woman relationships	159.96
1808	Gertrude of Wyoming	Campell	Wyoming Massacre (1778)	974.8
1809	Knickerbocker's history of New York	Irving	conduct of life	159.9
1810	The scottish chiefs	Porter	Revolutions	333
1811	The asylum, or Alonzo and Melissa	Mitchel	Man-woman relationships	159.96
1812	Childe Harold	Byron	Christian life	248.4896
1813	Bride of Abydos	Byron	Fore-edge painting	96.1
1814	Waverley	Scott	Revolutions	333
1815	Guy Mannering	Scott	Inheritance and succession	347.6
1816	Tales of my landlord	Scott	Revolutions	333
1817	Lalla Rookh	Moore	Fore-edge painting	96.1
1818	The heart of Midlothian.	Scott	revolutions	333
1819	The sketch book	Irving	conduct of life	159.9
1820	The abbott	Byron	Politics and government	321
1821	The spy	Cooper	spies	327.5
1822	Bracebridge Hall	Irving	conduct of life	159.9
1823	The pilot	Cooper	American civil war	930.365
1824	A narrative of the life of Mrs. Mary Jemison	Seaver	Abduction	343.6
1825	Lionel Lincoln	Cooper	American civil war	930.365
1826	The last of the Mohicans	Cooper	pioneer life	325.14
1827	The prairie	Cooper	pioneer life	325.14
1828	Lucy Temple	Rowson	Runaway teenagers	362.74
1829	The conquest of Granada	Irving	Spanish Conquest of Granada (Reino : 1476-	972.8515
1831	The Dutchman's fireside	Paulding	conduct of life	159.9
1832	The Alhambra	Irving	conduct of life	159.9
1833	he life and writings of major Jack Downing	Smith	Politics and government	321
1834	Outre-mer	Longfellow	Europe	914
1835	A tour on the prairies	Irving	pioneer life	325.14
1836	Astoria	Irving	pioneer life	325.14
1837	Nick of the woods	Bird	pioneer life	325.14
1838	The Robber	James	robbery	343.3
1839	The green mountain boys	Thompson	American civil war	930.365
1840	The old curiosity shop	Dickens	conduct of life	159.9
1841	The ancient regime	James	revolution	333
1842	American notes	Dickens	conduct of life	159.9
1843	Ned Myers	Cooper	Seafaring life	910.45
1844	The monks of monk hall	Lippard	conduct of life	159.9
1845	The smuggler	James	smuggling	364.133
1846	Napoleon and his marshalls	Headly	war	327.6
1847	Jane Eyre	Bronte	Governesses	371.1
1848	Vanity fair	Thackeray	Man-woman relationships	159.96
1849	Kavanagh	Longfellow	Man-woman relationships	159.96
1850	The scarlet letter	Hawthorne	Hypocrisy	159.98
1851	The golden legend	Longfellow	Leprosy–Patients	362.196
1852	Uncle Tom's cabin	Stowe	Plantation life	950
1853	Heir of Redcliffe	Yonge	conduct of life	159.9
1854	Hard times	Dickens	Social problems	301.153
1855	The song of Hiawatha	Longfellow	Race relations	301.183
1856	John Halifax, gentleman	Mulock	Industrial revolution	330.942
1857	Barchester Towers	Trollope	conduct of life	159.9
1858	bay path	Holland	pioneer life	325.14
1859	The Virginians	Thackeray	French and Indian War (United States : 1754-1763)	
1860	Malaeska, or the indian wife of the white hunter	Stephens	Man-woman relationships	159.96
1861	Great expectations	Dickens	Man-woman relationships	159.96
1862	Maum Guinea and her plantation children	Victor	Plantation life	950

TABLE II: Best sold novels 1863 – 1964

1863	History of the great rebellion	Ketell	war	327.6
1864	the American conflict	Greeley	war	327.6
1865	Our mutual friend	Dickens	Inheritance and succession	347.6
1866	Griffith Gaunt	Reade	Jealousy	159.983
1867	Beyound the Mississippi	Richardson	Travel	301.188
1868	The moonstone	Collins	Larceny	343.7
1869	Innocents abroad	Twain	Travel	301.188
1870	The heathen chinee	Harte	International relations	301.29511
1871	The hoosier schoolmaster	Egglestone	conduct of life	159.9
1872	Roghing it	Twain	Travel	301.188
1873	The gilded age	Twain	Political corruption	364.16
1874	Far from the madding crowd	Hardy	Man-woman relationships	159.96
1875	Opening a chestnut burr	Roe	conduct of life	159.9
1876	Tom Swayer	Twain	Race relations	301.183
1877	Self-raised	Southworth	Christian life	248.4896
1878	The leavenworth case	Green	murder	343.91
1879	Progress and poverty	George	depressions	930.33
1880	Ben-Hur	Wallace	roman empire	930
1881	Uncle Remus	Harris	Race relations	301.183
1882	The prince and the pauper	Twain	Impostors and imposture	364.163
1883	Treasure Island	Stevenson	Treasure hunterss	622.19
1884	The adventures of Huckleberry Finn	Twain	Race relations	301.183
1885	King Solomon's mines	Haggard	mines and mining	301.188
1886	The mayor of Casterbridge	Hardy	Man-woman relationships	159.96
1887	Mr. Barnes of New York	Gunter	Travel	301.188
1888	A romance of two worlds	Corelli	extraterrestrial life	52
1889	A connecticut yankee in King Arthur's court	Twain	magic	155.84
1890	A cigarette-maker's romance	Crawford	Man-woman relationships	159.96
1891	Colonel Carter of Carterville	Smith	reaction & reconstruction	320.9
1892	Don Orsino	Crawford	Political corruption	323
1893	The heavenly twins	Grand	Feminist fiction	301.183
1894	Coin's financial school	Harvey	Depressions	337
1895	Beside the Bonnie Brier Bush	Maclaren	Young men	343.81
1896	Tom Grogan	Smith	conduct of life	159.9
1897	Quo Vadis	Sienkiewiz	roman empire	930
1898	Caleb West, master diver.	Hopkinson !	divers	610.3
1899	David Harum: a story of American life.	Westcott	bankers	351.72
1900	To Have and to Hold	Johnston	pioneer life	325.14
1901	The Crisis	Churchill	American civil war	930.365
1902	The Viginian : a horseman of the plains	Wister	cattle stealing	930.33
1903	Lady Rose's Daughter	Ward	Illegitimate children	341.23
1904	The Crossing	Churchill	pioneer life	325.14
1905	The Marriage of William Ashe	Ward	mariage	340.2
1906	Coniston	Churchill	politics	327.1
1907	The Lady of the Decoration	Little	Russo-Japanese War (1904-1905)	930.365
1908	Mr. Crewe's Career	Churchill	Railroads	323
1909	The Inner Shrine: a novel of today	King	Women--Social life and customs	301.185
1910	The Rosary	Barclay	Women--Social life and customs	301.185
1911	The Broad Highway	Farnoll	Young men	343.81
1912	The Harvester	Stratton Po	Man-woman relationships	159.96
1913	The Inside of the Cup	Churchill	Hypocrisy	159.98
1914	The Eyes of the World	Bell Wright	Integrity	159.97
1915	The Turmoil	Tarkington	Integrity	159.97
1916	Seventeen : a tale of youth and summer time and the Baxter fa	Tarkington	Man-woman relationships	159.96
1917	Mr. Britling Sees It Through	Wells	World War (1914-1918)	930.365
1918	The U.P. Trail	Grey	Railroads	930.3
1919	The Four Horseman of the Apocalypse	Blasco	war	327.6
1920	The Man of the Forest	Grey	pioneer life	325.14
1921	Main Street & Babbitt	Sinclair	Married women	301.185
1922	If Winter Comes	Hutchinson	Marriage	930.347
1923	Black Oxen	Atherton	Women--Social life and customs	301.185
1924	So Big	Ferber	Feminist fiction	301.183

1925	Soundings	Gibbs	Women--Conduct of life	301.185
1926	The Private Life of Helen of Troy	Erskine	Helen of Troy	938
1927	Elmer Gantry	Sinclair	Hypocrisy	159.98
1928	The Bridge of San Luis Rey	Wilder	Fate and fatalism	321
1929	All Quiet on the Western Front	Remarque	World War (1914-1918)	930.365
1930	Cimarron	Ferber	Feminist fiction	301.183
1931	The Good Earth	Buck	Farmers' spouses	301.185
1932	The Good Earth	Buck	Farmers' spouses	301.185
1933	Anthony Adverse	Allen	Illegitimate children	341.23
1934	Anthony Adverse	Allen	Illegitimate children	341.23
1935	Green Light	Douglas	Conduct of life	159.9
1936	Gone with the Wind	Mitchell	Farmers' spouses	301.185
1937	Gone with the Wind	Mitchell	Farmers' spouses	301.185
1938	The Yearling	Kinnan Raw	Human-animal relationships	327.6
1939	The Grapes of Wrath	Steinbeck	Depressions	337
1940	How Green Was My Valley	Llewellyn	family relations	301.188
1941	The Keys of the Kingdom	Cronin	catholic missions	266.27
1942	The Song of Bernadett	Werfel	Christian saints	273
1943	The Robe	Douglas	Holy Coat	133.5
1944	Strange Fruit	Smith	Race relations	301.183
1945	Forever Amber	Cronin	Mistresses	930.32
1946	The King's General	Du Maurier	Civil War (Great Britain : 1642-1649)	930.3
1947	The Miracle of the Bells	Russel	movie stars	791.43028
1948	The Big Fisherman	Douglas	early church	930.22
1949	The Egyptian	Waltari	medicine	501
1950	The Cardinal	Robinson	Cardinals (clergy)	340.51
1951	From Here to Eternity	Jones	Armed Forces--Military life	327.6
1952	The Silver Chalice	Costain	Grail	133.5
1953	The Robe	Douglas	Holy Coat	133.5
1954	Not as a Stranger	Thompson	Man-woman relationships	159.96
1955	Marjorie Morningstar	Wouk	Jewish women	159.92
1956	Don't Go Near the Water	Brinkley	Armed Forces--Military life	327.6
1957	By Love Possessed	Cozzens	Man-woman relationships	159.96
1958	Doctor Zhivago	Pasternak	Revolution (Soviet Union : 1917-1921)	323.27
1959	Exodus	Uris	Emigration and immigration	325.17
1960	Advise and Consent	Drury	Politics and government	321
1961	The Agony and the Ecstasy	Stone	italian art	709.2
1962	Ship of Fools	Porter	microsociology	301
1963	The Shoes of the Fisherman	West	Arbitration (International law)	341.62
1964	The Spy Who Came in from the Cold	le Carre	cold war	930.3

TABLE III Best sold novels 1965 - 2015

1965	The Source	Michener	archeology	7.02
1966	Valley of the Dolls	Susann	movie industry	301.188
1967	The Arrangement	Kazan	advertsement & society	301.182
1968	Airport	Hailey	aviation industry	383
1969	Portnoy's Complaint	Roth	sexual behavior	159.92
1970	Love Story	Segal	Man-woman relationships	159.96
1971	Wheels	Blatty	auto industry	338.9
1972	Jonathan Livingston Seagull	Bach	spirituality	159.9
1973	Jonathan Livingston Seagull	Bach	spirituality	159.9
1974	Centennial	Michener	pioneer life	325.14
1975	Ragtime	Doctorow	Race relations	301.183
1976	Trinity	Uris	revolution	321.3
1977	The Silmarillion	Tolkien	Imaginary wars and battles	159.9
1978	Chesapeake	Michener	pioneer life	325.14
1979	The Matarese Circle	Ludlum	spies	327.5
1980	The Covenant	Michener	Race relations	301.183
1981	Noble House	Clavell	buisiness men	340.14
1982	E.T., The Extraterrestrial	Kotzwinkle	extraterrestrial life	52
1983	Return of the Jedi	Kahn	Imaginary wars and battles	159.9
1984	The Talisman	King	Imaginary wars and battles	159.9
1985	The Mammoth Hunters	Auel	Man-woman relationships	159.96
1986	It	King	Imaginary wars and battles	159.9
1987	The Tommyknockers	King	Imaginary wars and battles	159.9
1988	The Cardinal of the Kremlin	Clancy	spies	327.5
1989	Clear and Present Danger	Clancy	war on drugs	343.9
1990	The Plains of Passage	Auel	Man-woman relationships	159.96
1991	Scarlett	Ripley	Man-woman relationships	159.96
1992	Dolores Claiborne	King	murder	343.91
1993	The Pelican Brief	Grisham	murder	343.91
1994	The Rainmaker	Grisham	Insurance fraud	347.764
1995	The Runaway Jury	Grisham	Tobacco industry	338.93
1996	The Partner	Grisham	Larceny	343.7
1997	The Partner	Grisham	Larceny	343.7
1998	The Street Lawyer	Grisham	Homeless persons	301.181
1999	The Testament	Grisham	Inheritance and succession	347.6
2000	The Brethren	Grisham	Prisoners	343.81
2001	Desecration	Jenkins	Antichrist	14
2002	The Summons	Grisham	Inheritance and succession	347.6
2003	Harry Potter and the Order of the Phoenix	Rowling	magic	133.4
2004	The Da Vinci Code	Brown	Grail	133.5
2005	The Broker	Grisham	spies	327.5
2006	For One More Day	Albom	Suicidal behavior	362.6
2007	Harry Potter and the Deathly Hallows	Rowling	magic	133.4
2008	The Appeal	Grisham	environment	343.3
2009	The Lost Symbol	Brown	Freemasonry	133.4
2010	The Girl Who Kicked the Hornet's Nest	Larson	Political corruption	323
2011	The Litigators	Grisham	Products liability–Drugs	351.82
2012	Fifty Shades of Grey	James	Man-woman relationships	159.96
2013	Diary of a Wimpy Kid: Hard Luck	Kinney	Friendship	159.92
2014	The Fault in Our Stars	Green	Cancer in adolescence	159.92
2015	Go Set a Watchman	Harper	Race relations	301.183

TABLE IV; Best sold non fiction 1917 – 2012.

1917	Rhymes of a Red Cross Man by Robert W. Service	811.52
1918	Rhymes of a Red Cross Man by Robert W. Service	811.52
1919	The Education of Henry Adams by Henry Adams*	973.07202
1920	Now It Can Be Told by Philip Gibbs	921
1921	The Outline of History by H.G. Wells	909
1922	The Outline of History by H.G. Wells	909
1923	Etiquette by Emily Post	395
1924	Diet and Health by Lulu Hunt Peters	612.39
1925	Diet and Health by Lulu Hunt Peters	612.39
1926	The Man Nobody Knows by Bruce Barton	232.9
1927	The Story of Philosophy by Will Durant	190
1928	Disraeli by Andre Marouis	941.081
1929	The Art of Thinking by Ernest Dimnet	701.17
1930	The Story of San Michele by Axel Munthe	610.69
1931	Education of a Princess by Grand Duchess Marie	920.7
1932	The Epic of America by James Truslow Adams	973
1933	Life Begins at Forty by Walter B. Pitkin	895.136
1934	While Rome Burns by Alexander Woollcott	818.5
1935	North to the Orient by Anne Morrow Lindbergh	950
1936	Man the Unknown by Alexis Carrell	572
1937	How To Win Friends and Influence People by Dale Car	158.1
1938	The Importance of Living by Lin Yutang	170
1939	Days of Our Years by Pierre van Paassen	286.4
1940	I Married Adventure by Osa Johnson 923.9	923.9
1941	Berlin Diary by William L. Shirer	940.5343
1942	See Here, Private Hargrove	827.911
1943	Under Cover by John Roy Carlson	930.3
1944	I Never Left Home by Bob Hope	940.5496
1945	Brave Men by Ernie Pyle	940.542
1946	The Egg and I by Betty McDonald	630.1
1947	Peace of Mind by Johsua L. Liebman	150.13
1948	Crusade in Europe by Dwight D. Eisenhower	327.6
1949	White Collar Zoo by Clare Barnes Jr.	301.188
1950	Betty Crocker's Picture Cook Book by Betty Crocker	641.5
1951	Look younger, Live Longer by Gayelord Hauser	613.2
1952	The Holy Bible: Revised Standard Version	220.52
1953	The Holy Bible: Revised Standard Version	220.52
1954	The Holy Bible: Revised Standard Version	220.52
1955	Gift from the Sea by Anne Morrow Lindbergh	128
1956	Arthritis and Common Sense by Dan Dale Alexander	616.722
1957	Kids Say the Darndest Things! by Art Linklater	791.4
1958	Kids Say the Darndest Things! by Art Linklater	791.4
1959	The Autobiography of Mark Twain by Mark Twain)	928.1
1960	Folk Medicine by D.C. Jarvis	616.02
1961	The New English Bible: The New Testament	220.52
1962	Calories Don't Count by Dr. Herman Taller	615.854
1963	Happiness Is a Warm Puppy by Charles M. Schultz	741.5973
1964	Four Days by American Heritage and United Press Int	923.173
1965	How To Be a Jewish Mother by Dan Greenburg	817.54
1966	The Art of Memory by Frances A. Yates	808.5
1967	Death of a President by William Manchester	973.92209
1968	Better Homes and Gardens New Cook Book	641.5
1969	American Heritage Dictionary of the English Language	423
1970	Everything You Always Wanted To Know About Sex t	301.41763
1971	The Sensuous Man by "M"	301.41763
1972	Great Bridge by David McCullough)	624.23097
1973	The Joy of Sex by Alex Comfort (Crown)	613.96
1974	The Total Woman by Maribel Morgan	646.75
1975	Angels: God's Secret Agents by Billy Graham	235.3
1976	The Final Days by Bob Woodward and Carl Bernstein	364.1321
1977	Roots by Alex Haley	301.183
1978	If Life Is a Bowl of Cherries--What Am I Doing in the Pits? by Erma Bomt	818.5407
1979	Aunt Erma's Cope Book by Erma Bombeck	814.5407
1980	Crisis Investing: Opportunities and Profits in the Coming Great Depressic	332.678
1981	The Beverly Hills Diet by Judy Mazel	613.25
1982	Jane Fonda's Workout Book by Jane Fonda	613.7045
1983	In Search of Excellence: Lessons from America's Best-Run Companies t	658.00973
1984	Iacocca: An Autobiography by Lee Iacocca with William Novak	338.76292
1985	Iacocca: An Autobiography by Lee Iacocca with William Novak	338.76292
1986	Fatherhood by Bill Cosby	159.92
1987	Time Flies by Bill Cosby	612.67
1988	The 8-Week Cholesterol Cure by Robert E. Kowalski	616.12307
1989	All I Really Need To Know I Learned in Kindergarten: Uncommon Though	812
1990	A Life on the Road by Charles Kuralt	917.30492
1991	Me: Stories of My Life by Katharine Hepburn	791.43028
1992	The Way Things Ought To Be by Rush Limbaugh	818.54
1993	See, I Told You So by Rush Limbaugh	306.0973
1994	Operating system; a journal of my son's first year	306.8
1995	My American Journey by Colin L. Powell	305.3
1996	Make the Connection by Oprah Winfrey	613.7
1997	Angela's Ashes by Frank McCourt	454
1998	The 9 Steps to Financial Freedom by Suze Orman	332.024
1999	Tuesdays with Morrie by Mitch Albom	378.12092
2000	Who Moved My Cheese? by Spencer Johnson	158.1
2001	The Prayer of Jabez by Bruce Wilkinson	242.722
2002	Self Matters by Dr. Phil McGraw	158.1
2003	The Purpose-Driven Life by Rick Warren	248.4
2004	The Purpose-Driven Life by Rick Warren	248.4
2005	Natural Cures "They" Don't Want You to Know About by Kevin Trudeau	615.535
2006	The Innocent Man by John Grisham	345.76603
2007	The Secret by Rhonda Byrne	131
2008	The Last Lecture by Randy Pausch	158.1
2009	Going Rogue: An American Life by Sarah Palin	973.93109
2010	Decision Points by George W. Bush	973.93109
2011	Steve Jobs by Walter Isaacson	621.39092
2012	No Easy Day by Mark Owen	958.10461

TABLE V: Dewey Decimal Categories

Dewey Decimal Categories

by broad topics (hundreds) and more specific topics (tens)

000 Generalities
- 010 Bibliography
- 020 Library & information sciences
- 030 General encyclopedic works
- 040 Special topics
- 050 General serials & their indexes
- 060 General organizations & museums
- 070 New media, journalism, publishing
- 080 General collections
- 090 Manuscripts & rare books

100 Philosophy & psychology
- 110 Metaphysics
- 120 Epistemology, causation, humankind
- 130 Paranormal phenomena
- 140 Specific philosophical schools
- 150 Psychology
- 160 Logic
- 170 Ethics (moral philosophy)
- 180 Ancient, medieval, oriental philosophy
- 190 Modern western philosophy

200 Religion
- 210 Natural theology
- 220 Bible
- 230 Christian theology
- 240 Christian moral & devotional theology
- 250 Christian orders & local churches
- 260 Christian social theology
- 270 Christian church history
- 280 Christian denominations & sects
- 290 Other & comparative religions

300 Social Science
- 310 General statistics
- 320 Political science
- 330 Economics
- 340 Law
- 350 Public administration
- 360 Social problems & services
- 370 Education
- 380 Commerce, communications, transport
- 390 Customs, etiquette, folklore

400 Language
- 410 Linguistics
- 420 English & Anglo-Saxon languages
- 430 Germanic languages (German)
- 440 Romance languages (French)
- 450 Italian, Romanian, Rhaeto-Romanic
- 460 Spanish & Portuguese languages
- 470 Italic languages (Latin)
- 480 Hellenic languages (Classical Greek)
- 490 Other languages

500 Natural science & mathematics
- 510 Mathematics
- 520 Astronomy & allied sciences
- 530 Physics
- 540 Chemistry & allied sciences
- 550 Earth sciences
- 560 Paleontology & Paleozoology
- 570 Life sciences
- 580 Botanical sciences
- 590 Zoological sciences

600 Technology (applied sciences)
- 610 Medical sciences (Medicine, Psychiatry)
- 620 Engineering
- 630 Agriculture
- 640 Home economics & family living
- 650 Management
- 660 Chemical engineering
- 670 Manufacturing
- 680 Manufacture for specific use
- 690 Buildings

700 The arts
- 710 Civic & landscape art
- 720 Architecture
- 730 Sculpture
- 740 Drawings & decorative arts
- 750 Paintings & painters
- 760 Graphic arts (Printmaking & prints)
- 770 Photography
- 780 Music
- 790 Recreational & performing arts

800 Literature & rhetoric
- 810 American literature in English
- 820 English literature
- 830 Literature of Germanic language
- 840 Literatures of Romance language
- 850 Italian, Romanian, Rhaeto-Romanic Literatures
- 860 Spanish & Portuguese literatures
- 870 Italic literatures (Latin)
- 880 Hellenic literatures (Classical Greek)
- 890 Literatures of other languages

900 Geography & history
- 910 Geography & travel
- 920 Biography, genealogy, insignia
- 930 History of the ancient world
- 940 General history of Europe
- 950 General history of Asia (Far East)
- 960 General history of Africa
- 970 General history of North America
- 980 General history of South America
- 990 General history of other areas

TABLE VI; main protagonist types in novels 1863- 1963.

1863	History of the great rebellion	Ketell	statesmen	1131990	203.7582
1864	the American conflict	Greeley	slaves	1120522	201.69396
1865	Our mutual friend	Dickens	secretary	1204270	216.7686
1866	Griffith Gaunt	Reade	Young men-	1183266	212.98788
1867	Beyound the Mississippi	Richardson	traveler	1155558	208.00044
1868	The moonstone	Collins	Young men-	1183266	212.98788
1869	Innocents abroad	Twain	traveler	1155558	208.00044
1870	The heathen chinee	Harte	chinese	1033393	186.01074
1871	The hoosier schoolmaster	Egglestone	teachhers	1144248	205.96464
1872	Roghing it	Twain	traveller	1155558	208.00044
1873	The gilded age	Twain	politician	1204885	216.8793
1874	Far from the madding crowd	Hardy	Women farn	1177667	211.98006
1875	Opening a chestnut burr	Roe	Young men-	1183266	212.98788
1876	Tom Swayer	Twain	boys	1205150	216.927
1877	Self-raised	Southworth	Women autl	1177210	211.8978
1878	The leavenworth case	Green	lawyer	1204505	216.8109
1879	Progress and poverty	George	wealth	1172973	211.13514
1880	Ben-Hur	Wallace	soldier	1210272	217.84896
1881	Uncle Remus	Harris	Storytellers	1134165	204.1497
1882	The prince and the pauper	Twain	boys	1076481	193.76658
1883	Treasure Island	Stevenson	boys	1205150	216.927
1884	The adventures of Huckleberry Finn	Twain	boys	1205150	216.927
1885	King Solomon's mines	Haggard	traveller	1155558	208.00044
1886	The mayor of Casterbridge	Hardy	politician	1204623	216.83214
1887	Mr. Barnes of New York	Gunter	traveller	1155558	208.00044
1888	A romance of two worlds	Corelli	Women autl	1177210	211.8978
1889	A connecticut yankee in King Arthur's court	Twain	time travelle	1151176	207.21168
1890	A cigarette-maker's romance	Crawford	Young men-	1183266	212.98788
1891	Colonel Carter of Carterville	Smith	soldier	1204885	216.8793
1892	Don Orsino	Crawford	migrant worl	1204048	216.72864
1893	The heavenly twins	Grand	Women	1176568	211.78224
1894	Coin's financial school	Harvey	banker	1205293	216.95274
1895	Beside the Bonnie Brier Bush	Maclaren	college stuc	1206715	217.2087
1896	Tom Grogan	Smith	stevedore	1204597	216.82746
1897	Quo Vadis	Sienkiewiz	politician	1204885	216.8793
1898	Caleb West, master diver.	Hopkinson :	diver	1210280	217.8504
1899	David Harum: a story of American life.	Westcott	banker	1205293	216.95274
1900	To Have and to Hold	Johnston	Ranchers	1204597	216.82746
1901	The Crisis	Churchill	soldier	1204885	216.8793
1902	The Viginian : a horseman of the plains	Wister	policeman	1204583	216.82494
1903	Lady Rose's Daughter	Ward	secretary	1077602	193.96836
1904	The Crossing	Churchill	politician	1204604	216.82872
1905	The Marriage of William Ashe	Ward	politician	1204623	216.83214
1906	Coniston	Churchill	politician	1204318	216.77724
1907	The Lady of the Decoration	Little	politician	1204082	216.73476
1908	Mr. Crewe's Career	Churchill	Lawyer	1204318	216.77724
1909	The Inner Shrine: a novel of today	King	Young wom	1204155	216.7479
1910	The Rosary	Barclay	Artists	1219920	219.5856
1911	The Broad Highway	Farnoll	scholar	1212344	218.22192
1912	The Harvester	Stratton Poi	harvester	1204604	216.82872
1913	The Inside of the Cup	Churchill	clergy man	1204155	216.7479
1914	The Eyes of the World	Bell Wright	Artists	1202804	216.50472
1915	The Turmoil	Tarkington	Business m	1240052	223.20936
1916	Seventeen : a tale of youth and summer time and the Baxter fa	Tarkington	High school	1240052	223.20936
1917	Mr. Britling Sees It Through	Wells	writer	1212484	218.24712
1918	The U.P. Trail	Grey	Surveyors	1204583	216.82494
1919	The Four Horseman of the Apocalypse	Blasco	soldier	1204289	216.77202
1920	The Man of the Forest	Grey	Ranchers	1253965	225.7137
1921	Main Street & Babbitt	Sinclair	college stuc	1204560	216.8208
1922	If Winter Comes	Hutchinson	Unmarried n	1204155	216.7479
1923	Black Oxen	Atherton	Artists	1204155	216.7479
1924	So Big	Ferber	migrant worl	1204048	216.72864
1925	Soundings	Gibbs	Artists	1219920	219.5856

1926	The Private Life of Helen of Troy	Erskine	queens	1085637	195.41466
1927	Elmer Gantry	Sinclair	clergy man	1204155	216.7479
1928	The Bridge of San Luis Rey	Wilder	clergy man	1205190	216.9342
1929	All Quiet on the Western Front	Remarque	soldier	1210272	217.84896
1930	Cimarron	Ferber	Journalists	1205031	216.90558
1931	The Good Earth	Buck	farmer	1206073	217.09314
1932	The Good Earth	Buck	farmer	1206073	217.09314
1933	Anthony Adverse	Allen	merchant	1204155	216.7479
1934	Anthony Adverse	Allen	merchant	1204155	216.7479
1935	Green Light	Douglas	Physicians	1204604	216.82872
1936	Gone with the Wind	Mitchell	farmer	1204622	216.83196
1937	Gone with the Wind	Mitchell	farmer	1204622	216.83196
1938	The Yearling	Kinnan Raw	Boys	1205150	216.927
1939	The Grapes of Wrath	Steinbeck	Migrant agri	1204928	216.88704
1940	How Green Was My Valley	Llewellyn	Miners	1219671	219.54078
1941	The Keys of the Kingdom	Cronin	clergy man	1206073	217.09314
1942	The Song of Bernadett	Werfel	saints	1202813	216.50634
1943	The Robe	Douglas	soldier	1204885	216.8793
1944	Strange Fruit	Smith	Unmarried r	1204622	216.83196
1945	Forever Amber	Cronin	prostitute	1219920	219.5856
1946	The King's General	Du Maurier	aristocrats	1038255	186.8859
1947	The Miracle of the Bells	Russel	publicity ag	1204155	216.7479
1948	The Big Fisherman	Douglas	Christian sa	1204885	216.8793
1949	The Egyptian	Waltari	Physicians	1208755	217.5759
1950	The Cardinal	Robinson	clergy man	1204155	216.7479
1951	From Here to Eternity	Jones	Soldiers	1208724	217.57032
1952	The Silver Chalice	Costain	Artists	1204885	216.8793
1953	The Robe	Douglas	soldier	1204885	216.8793
1954	Not as a Stranger	Thompson	Physicians	1204155	216.7479
1955	Marjorie Morningstar	Wouk	Artists	1210280	217.8504
1956	Don't Go Near the Water	Brinkley	soldier	1242982	223.73676
1957	By Love Possessed	Cozzens	Lawyer	1204505	216.8109
1958	Doctor Zhivago	Pasternak	Physicians	1207312	217.31616
1959	Exodus	Uris	Holocaust s	1204236	216.76248
1960	Advise and Consent	Drury	politician	1204505	216.8109
1961	The Agony and the Ecstasy	Stone	artist	1204565	216.8217
1962	Ship of Fools	Porter	traveler	1155558	208.00044
1963	The Shoes of the Fisherman	West	clergy	1205190	216.9342
1964	The Spy Who Came in from the Cold	le Carre	spies	975843	175.65174

TABLE VII; Evolution of the printed book collection of the Library of Congress

1869	11,262	210,279	5.36%
1870	11,512	221,791	5.19%
1871	39,178	260,969	15.01%
1872	9,499	270,468	3.51%
1873	12,407	282,875	4.39%
1874	15,405	298,280	5.16%
1875	19,350	317,630	6.09%
1876	17,590	335,220	5.25%
1877	20,021	355,241	5.64%
1878	21,537	376,778	5.72%
1879	20,367	397,145	5.13%
1880	0	397,145	0.00%
1881	23,304	420,449	5.54%
1882	59,984	480,433	12.49%
1883	33,365	513,798	6.49%
1884	31,246	545,044	5.73%
1885	20,447	565,491	3.62%
1886	16,544	582,035	2.84%
1887	15,279	597,314	2.56%
1888	18,824	616,138	3.06%
1889	17,936	634,074	2.83%
1890	15,211	649,285	2.34%
1891	21,186	670,471	3.16%
1892	24,284	694,755	3.50%
1893	27,260	722,015	3.78%
1894	18,652	740,667	2.52%
1895	20,311	760,978	2.67%
1896	20,825	781,803	2.66%
1897	31,304	813,107	3.85%
1898	23,050	913,285	2.52%
1899	31,354	944,639	3.32%
1900	38,110	982,749	3.88%
1901	76,481	1,059,230	7.22%
1902	81,971	1,141,201	7.18%
1903	79,394	1,220,595	6.50%
1904	78,791	1,299,386	6.06%
1905	68,951	1,368,337	5.04%
1906		1,368,337	0.00%
1907	54,604	1,422,941	3.84%
1908	100,067	1,523,008	6.57%
1909	167,677	1,702,685	9.85%
1910	90,473	1,793,158	5.05%
1911	98,571	1,891,729	5.21%
1912	120,664	2,012,393	6.00%
1913	115,862	2,128,255	5.44%
1914	125,054	2,253,309	5.55%
1915	110,564	2,363,874	4.68%
1916	88,101	2,451,974	3.59%
1917	85,948	2,537,922	3.39%
1918	76,601	2,537,922	3.02%
1919	96,033	2,710,556	3.54%
1920	120,777	2,831,333	4.27%
1921	86,923	2,918,256	2.98%
1922	82,152	3,000,408	2.74%
1923	88,933	3,089,341	2.88%
1924	89,763	3,179,104	2.82%
1925	106,661	3,285,765	3.25%
1926	134,580	3,420,345	3.93%
1927	136,422	3,566,767	3.82%
1928	169,735	3,726,502	4.55%
1929	180,802	3,907,304	4.63%
1930	196,632	4,103,936	4.79%
1931	188,352	4,292,288	4.39%
1932	185,143	4,477,431	4.14%
1939	236,416	5,591,710	4.23%
1940	246,898	6,102,259	4.05%
1941	274,133	6,349,157	4.32%
1942	255,871	6,609,387	3.87%
1943	213,061	6,822,448	3.12%
1944	481,733	7,304,181	6.60%
1945	572,821	7,877,002	7.27%
1946	69,458	7,946,460	0.87%
1947	240,604	8,187,064	2.94%
1948	200,321	8,387,358	2.39%
1949	302,254	8,689,639	3.48%
1950	267,354	8,956,993	2.98%
1951	284,772	9,241,765	3.08%
1952	336,936	9,578,701	3.52%
1953		9,846,831	0.00%
1954	308,476	10,155,307	3.04%
1955	357,741	10,513,048	3.40%
1956	462,946	10,975,994	4.22%
1957	444,776	11,420,770	3.89%
1958	373,768	11,794,538	3.17%
1959	364,098	12,158,636	2.99%
1960	362,320	12,520,956	2.89%
1961	347,359	12,868,315	2.70%
1962	204,673	13,072,988	1.57%
1963	218,441	13,291,429	1.64%
1964	276,464	13,567,893	2.04%
1965		13,653,168	0.00%
1966	351,251	13,767,403	2.55%
1967		14,107,259	0.00%
1968	407,210	14,479,171	2.81%
1969		14,848,317	0.00%
1970	436,133	15,258,327	2.86%
1971		15,660,523	0.00%
1972	361,811	16,022,327	2.26%
1973		16,466,899	0.00%
1974	328,896	16,761,198	1.96%
1975	251,803	17,013,001	1.48%
1976	255,307	17,268,308	1.48%
1977	279,583	17,547,891	1.59%
1978	319,901	17,867,792	1.79%
1979	1,063,113	18,930,905	5.62%
1980	297,598	19,155,165	1.55%
1981		19,578,334	0.00%
1982	187,664	19,721,066	0.95%
1983	241,361	19,962,427	1.21%
1984	250,000	20,212,427	1.24%
1985	274,580	20,487,007	1.34%
1986	278,215	20,765,222	1.34%
1987	284,909	21,050,131	1.35%
1988	291,210	21,341,341	1.36%
1989	275,000	21,616,341	1.27%
1990	303,029	21,919,370	1.38%
1991	287,014	22,206,384	1.29%
1992	260,000	22,466,384	1.16%
1993	260,000	22,726,384	1.14%
1994	260,000	22,986,384	1.13%
1995	345,424	23,331,808	1.48%
1996	317,497	23,649,305	1.34%
1997	343,965	23,993,270	1.43%
1998	310,000	24,303,270	1.28%
1999	253,172	24,556,442	1.03%
2000	282,930	24,839,372	1.14%
2001	326,616	25,165,988	1.30%
2002	1,463,019	26,629,007	5.49%
2009	596,147	29,096,780	2.05%
2010	401,003	29,497,783	1.36%
2011	480,004	29,977,787	1.60%
2012	510,124	30,487,911	1.67%
2013	315,975	30,803,886	1.03%
2014	324,020	31,127,906	1.04%
Average annual growth			3.42%

Made in the USA
Monee, IL
05 July 2020

34514620R00121